MANUAL OF GRAPHIC TECHNIQUES 3

FOR ARCHITECTS, GRAPHIC DESIGNERS, & ARTISTS

TOM PORTER AND SUE GOODMAN

Charles Scribner's Sons · New York

Acknowledgments

The authors would like to thank the following people for their help in providing material and information for the production of this manual:

Roy Barris, Colin Catron, Ian Cole, Jack Foreman, Martin Gordon, Dennis Hall, Betty Hill, Gary Jemmett, John Kirby, David Koh, Jean Lodge, Pat Osborne, Roger Osborn-King, Nigel Somner, Kathleen Tramner.

Special thanks are also due to Iradj Parvaneh and Jon Davidson for the photography, to Pat McNiff for typing the text, and to Richard A. Abbott (Execuscope, Bournemouth, Dorset, England) for his expert advice and supply of visual material for the section on modelscope photography.

Library of Congress Cataloging in Publication Data

Porter, Tom.

Manual of graphic techniques 3.

Includes index.
1. Graphic arts--Techniques. I. Goodman, Sue.
II. Title. III. Title: Manual of graphic techniques three.
NC1000.P666 1983 741.6 83-14087
ISBN 0-684-18018-9

3 5 7 9 11 13 15 17 19 Q/P 20 18 16 14 12 10 8 6 4

Printed in the United States of America.

TABLE OF CONTENTS

Introduction

Manual of Graphic Techniques 3 has been developed specifically as a companion volume to its forerunners, Manual of Graphic Techniques 1 and 2. Based on the self-contained page layouts and step-by-step frames of information devised for its predecessors, this manual examines wider design techniques together with some disciplines often overlooked or ignored in basic design courses.

For example, as the appearance of student presentations is often marred by the quality of lettering, this manual opens with an introduction to a basic pen-lettered alphabet. Also studied is the basic construction of a sans-serif display alphabet and roman lettering, the latter providing the elegant standard against which all other letter forms are judged. Also, in surviving over two thousand years, roman lettering exists as an important link between written communication as a stone-carved medium and its adaption in modern print technology.

Hand lettering is followed by some basic principles of graphic reproduction and typography. These are included because of the shift of designers toward a proliferation of ideas through the print medium and--in a competitive climate--an increasing search for wider exposure. Also included in this section are hints and tips when preparing artwork for transformation by the reproduction camera, as well as layout methods and techniques for combining words and pictures in the printed format.

Another method of image transformation and proliferation is the more accessible processes of manual printmaking. These range from the immediacy of monoprinting and relief printing to basic procedures of linocutting, screenprinting, and lithography. These occupy the next section. The rudiments of each process are described for the beginner.

The application of printmaking in the creation of models is well known. However, it is the recent resurgence of modelmaking as a means of realizing design ideas "in the round" that is considered an important trend by the authors. The concluding section therefore examines the various methods and materials of model construction, their roles in design and, via the introduction of the camera and modelscope, how they can be translated into photographs conveying powerful illusions of reality.

1 LETTERING DESIGN

Basic Freehand Pen Lettering

1

Very often, hand lettering is used to annotate, identify, and clarify information that is carried in a design drawing. It is also used because it is sometimes faster, more convenient, and more economical than drawing by mechanical means, such as instant lettering and stencils. However, when hand-lettered information is badly formed, it can appear the weakest element in a design presentation, with its malformation distracting the viewer from the message it conveys.

As the central function of lettering is that of communicating information quickly, it is crucial that the designer develop a legible style. For example, in architectural design, production drawings usually include written notes that communicate directly with those who will construct the building. In this context, hand lettering will be hampered by clever or overstylized treatments. Therefore, the simpler and clearer the formation of letters and words, the more legible the means of communication.

Although the following alphabet is based on the use of a lettering pen, it is a good idea for the beginner to approach hand lettering as an extension of normal handwriting. Initially, a pencil, fountain pen, or technical pen can be used in the practice of basic letter formation. Once a degree of control has been established, the lettering pen technique should then be attempted.

ABCDEFGHIJKLMNOPQRS
TUVWXYZ 1234567890
abcdefghijklmnopqrstuvwxyz

N.B.: Don't worry too much about early mistakes. Practice sessions aiming at the achievement of individually proportioned letters should progress to speed tests in which letters are drawn with the minimum number of strokes.

2

The following alphabet results from the use of a lettering pen such as a Graphos, and a chisel- or square-cut nib.

First, remove any protective lacquer from a new nib by immersing it briefly in boiling water. Pens not fitted with an integrated ink-feeding system are loaded by drawing an ink-loaded brush over their up-turned reservoir.

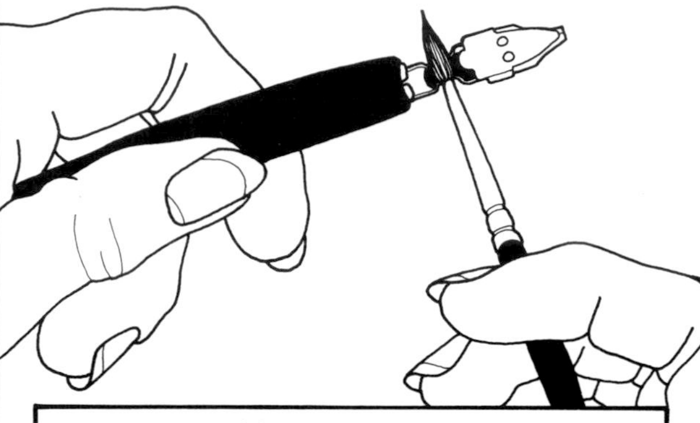

3

Throughout the drawing action, the pen should be held in a relaxed manner and at a constant angle of 45 degrees to the line of lettering, with the entire edge of the nib in contact with the paper.

N.B.: Avoid working on textured paper, as this can both inhibit spontaneity and clog the nib.

4

An oblique-cut nib is available for left-handed calligraphers. However, in both left- and right-handed drawing positions, a comfortable drawing hand position should be found, with the unused fingers curled to act as a rest on the paper.

5

As an aid to accuracy, beginners should use lightly penciled guidelines. These should be of a height equal to seven or eight times the thickness of the lettering nib in use.

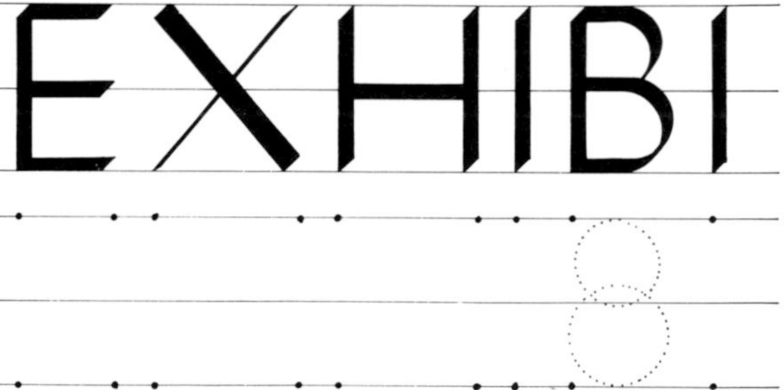

A further aid is preliminary plotting of proportions using penciled dots. However, as the ultimate goal is a spontaneously formed letter, this practice should be abandoned when confidence is gained.

Basic Freehand Pen Lettering

A trial exercise in forming various strokes will develop sensitivity toward the mark-making ability of the pen as well as find the best position in which to hold it. Initial experiments in making continuous strokes should be followed by exercises in forming individual strokes that practice the presentation and removal of the nib in a clean, decisive manner.

N.B.: Make deliberate strokes, always drawing toward yourself. Never exert pressure--allow the pen to do the work.

7

Once you have established a degree of familiarity with the pen, attempt the alphabet. Aim for simple, well-formed letters, remembering to draw them by using the slightest of pressure and holding the pen at a constant 45-degree angle. Also, aim to make each letter form distinct, with no chance of its being mistaken for another.

Each of the above capital and lowercase letters is shown with a suggested order and direction of pen-stroke. Some letters can be formed in several ways. After much practice you will be able to form them using fewer strokes.

N.B.: This alphabet can be drawn using a technical pen or a pencil, in which case the drawing instrument should be held in the upright position.

A Sans Serif Alphabet

1

The construction of the following alphabet is based on the proportion of the square. It is sans serif, that is, without serifs, and can be quickly drafted using a T-square, set square, and a pair of compasses with pencil or pen. Its main construction is based generally on roman lettering. This accounts for adjustments to its relationship with the square as reductions in the overall width of some of its full letters respond to the existence of serifs in the original. The thickness of the strokes is shown as one-ninth that of the letter height. However, depending on its design application, letter thickness can be reduced for delicacy or made heavier or bolder for added emphasis.

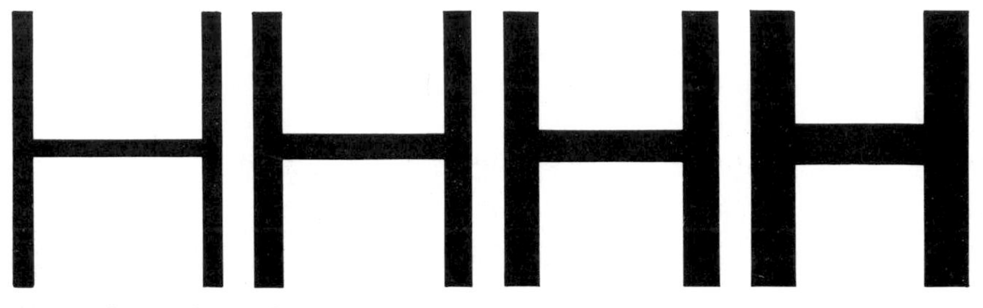

N.B.: When making letters bolder than those shown here, the extra thickness should be added on the inside of the strokes.

2

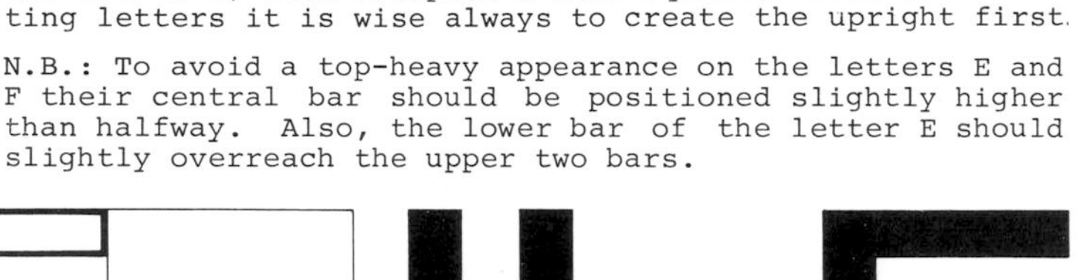

The letters I, L, F, and E are the simplest forms to construct. Apart from the upright stroke that represents the letter I, each occupies a half square. When constructing letters it is wise always to create the upright first.

N.B.: To avoid a top-heavy appearance on the letters E and F their central bar should be positioned slightly higher than halfway. Also, the lower bar of the letter E should slightly overreach the upper two bars.

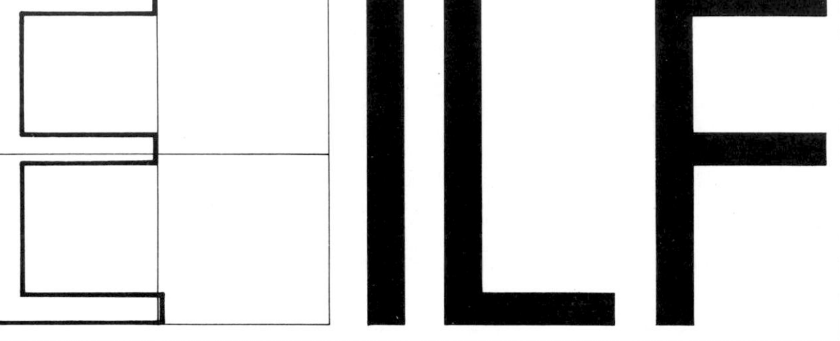

The letter T fills the square, its crossbar being reduced to four-fifths the width of the square to avoid top-heaviness.

N.B.: Another optical refinement is a slight reduction of thickness in the crossbar, which can be applied to all horizontal strokes in the alphabet.

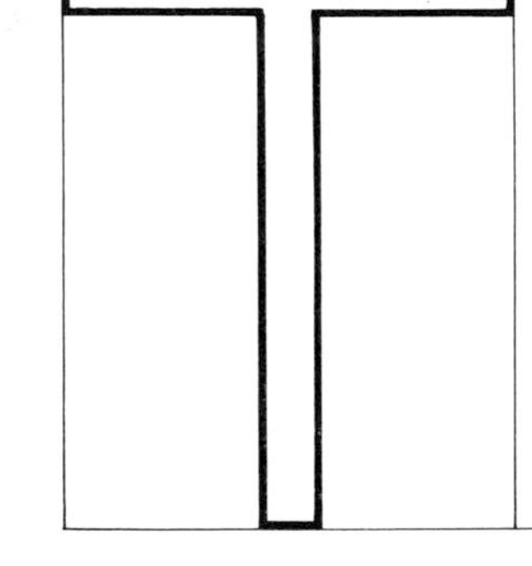

The letter H fills the square, its overall width reduced to four-fifths the width of the square. Apart from the slight reduction to its thickness, the crossbar should be positioned slightly higher than halfway to compensate for an optical effect of top-heaviness if the crossbar is drawn centrally.

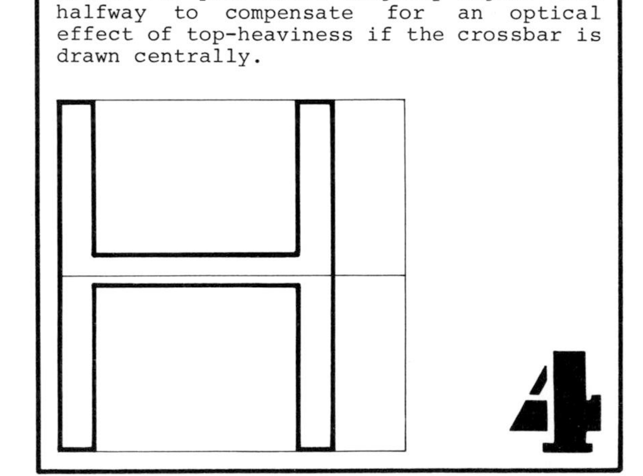

4

5

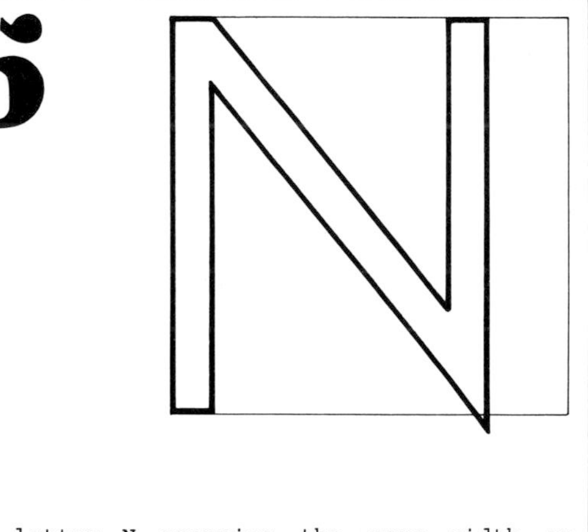

The letter N occupies the same width as the letter H. Notice that the upper intersection of the diagonal is blunted while the lower intersection is pointed--the latter sitting just below the baseline.

A Sans Serif Alphabet

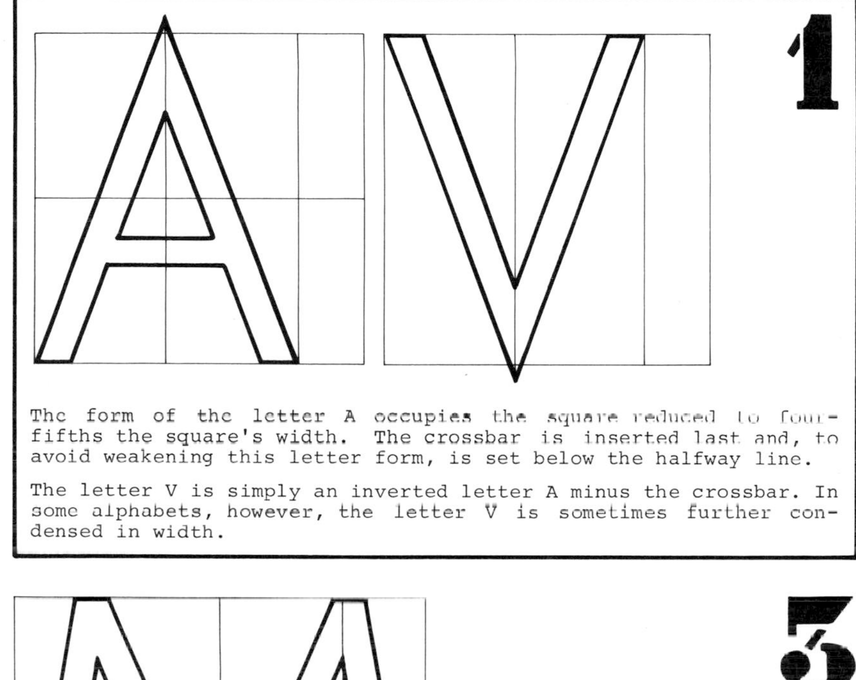

The form of the letter A occupies the square reduced to four-fifths the square's width. The crossbar is inserted last and, to avoid weakening this letter form, is set below the halfway line.

The letter V is simply an inverted letter A minus the crossbar. In some alphabets, however, the letter V is sometimes further condensed in width.

The letter M also fills the 1 1/4-square format, reversing that of the letter W but with a different optical adjustment. In this letter the central, inverted triangular space is drawn as slightly larger than its outer counterparts, which have narrower baselines.

N.B.: This letter can also be made to fit the square, but this contraction necessitates the introduction of rather ugly, shortened center strokes to avoid an appearance of being overcrowded.

In its double role when forming the letter W, the V does become condensed, with the two V's occupying 1 1/4 squares. A proportional adjustment should be made when constructing this letter so that the central, triangular space is made slightly smaller than its outer and inverted counterparts. This optical adjustment is achieved by giving the central triangle a narrower base when constructing the two outer strokes.

N.B.: This letter can be constructed with pointed intersections if required. When they are used, however, the points should sit just above and below the line.

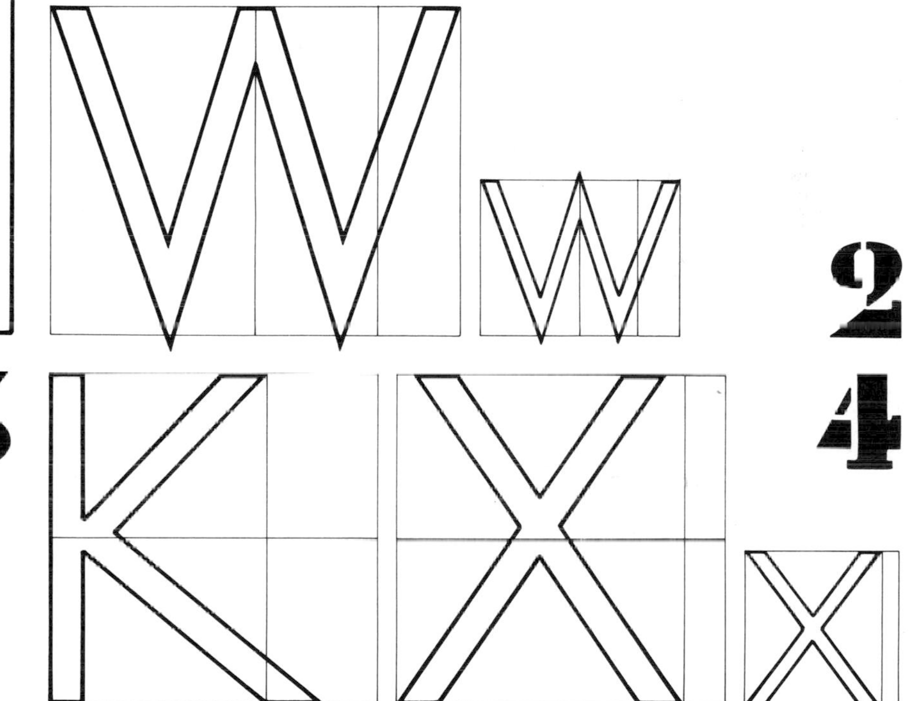

The upper arm of the letter K extends approximately two-thirds the width of the square. Draw the upright first and then add the upper diagonal stroke so that its left-hand edge penetrates halfway into the upright on the center line. An optical adjustment that makes this form appear more balanced is to allow the lower stroke to overreach the limits of its overhead counterpart slightly.

The letter X occupies the square minus the width of one stroke. Its center of balance can be raised to avoid a squat appearance by slightly reducing the upper triangular space.

A Sans Serif Alphabet

1

The arms of the letter Y extend to within two stroke widths of the full square. Construct the inverted triangle first so that the arms make their connection with the upright just below the halfway line.

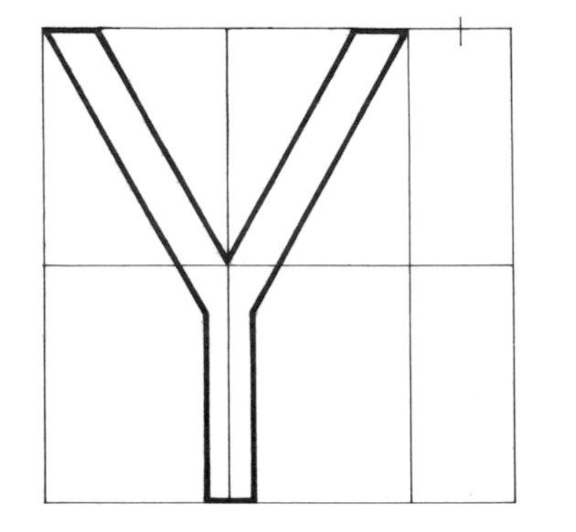

The letter Z extends to four-fifths the width of the square. First construct the diagonal, then add the horizontal bars. Top-heaviness is avoided by slightly shortening the upper bar.

2

The letter J fills just over half a square. Follow the construction of the loop with that of the upright, the former being drawn from the center of the lower quarter square.

N.B.: Be careful to avoid ugly junctions between curve and upright during construction.

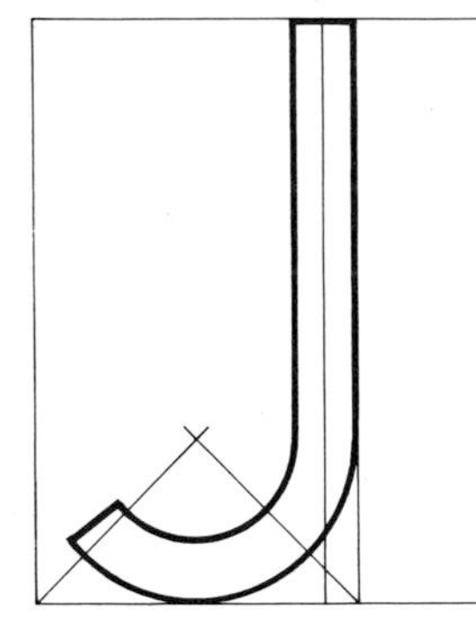

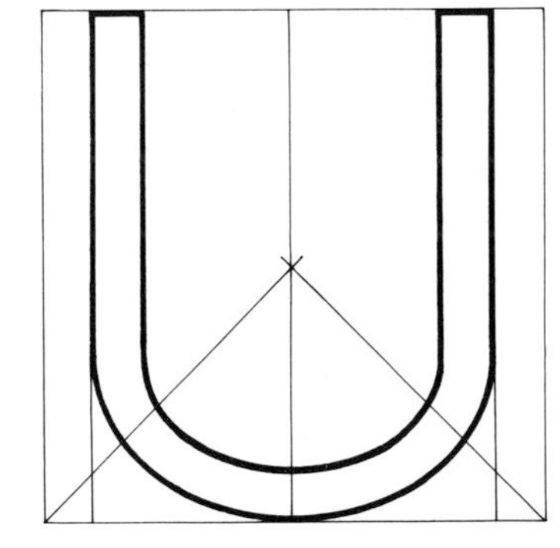

The letter U occupies the square reduced by one-fifth of its width. Once again, it is better to draw the curve first and then add the two upright strokes.

The letters C, Q, and G are all relatives of the letter O and each occupy the full square. If required, the upper and lower curves of the letters C and G can be softened as shown by the dotted lines. The letter G can be constructed from a variety of extensions, each used as a means of eliminating any confusion with the letter C.

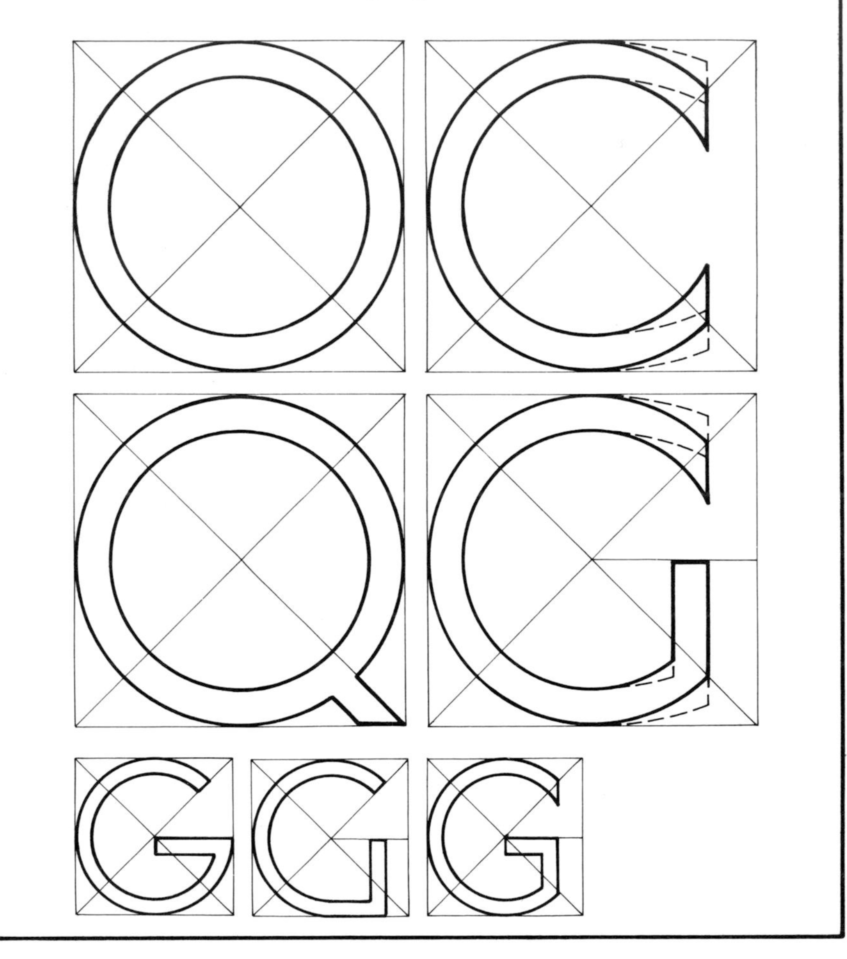

A Sans Serif Alphabet

The letter D fills the square minus the width of one stroke. First construct the upright. The center of the square locates the semi-circles that, when inscribed, merge the curve into the upper and lower horizontal sections.

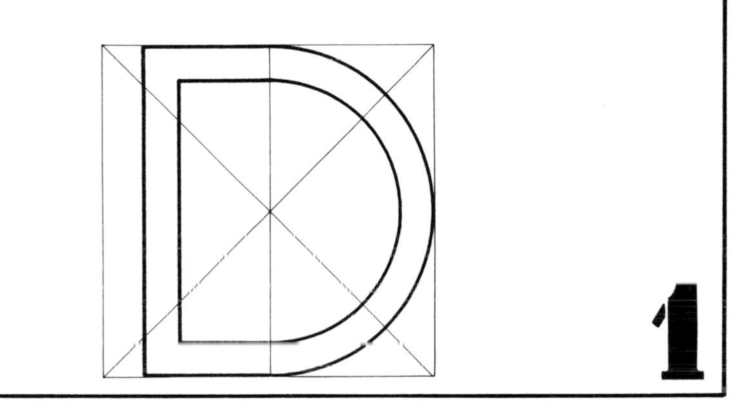

1

2

The letters P and R have a common loop whose belly extends out just beyond the half square, its lower section connecting back with the upright slightly below the halfway line. The inside edge of the letter R's diagonal leg begins on the loop's underside at a point immediately below the center for the semicircles.

N.B.: The diagonal leg extends out beyond the loop by approximately the thickness of its stroke.

If constructed of two equally sized loops, the letter B can appear unbalanced and top-heavy. In order to compensate for this optical illusion, the upper loop is reduced in size to allow a larger, lower loop that extends out just beyond the limits of the upper loop.

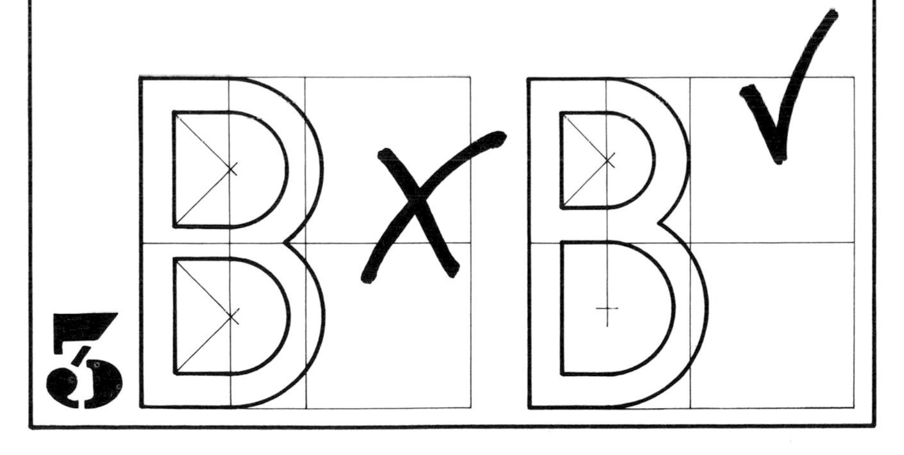

3

4

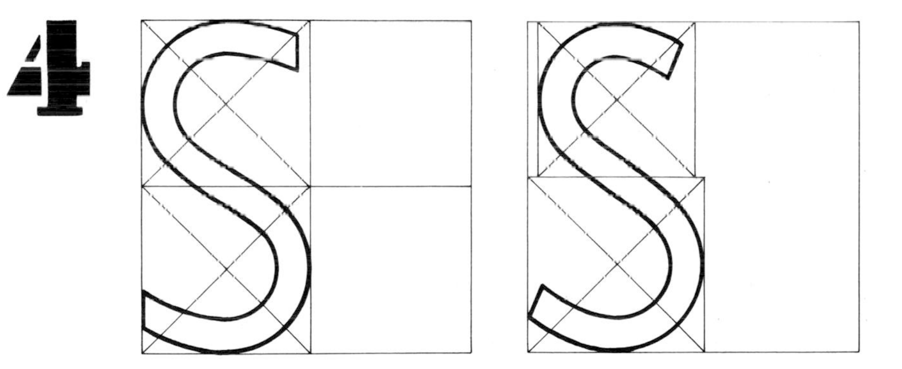

The letter S is formed by two circles occupying a half square. Again, rigid use of these guidelines to form this letter can result in a mechanical appearance not in character with this letter form.

It is better to use the basic construction as a framework on which to build up this letter. In doing so, it should not be necessary to make part of the line straight when changing from one circle to the other. Each end should break away slightly from the circle and follow a graceful line that allows the lower section to be slightly larger than the upper one.

Roman Lettering

1

Roman lettering provides a most beautiful alphabet whose formation comprises optically adjusted and elegant forms. The following pages explain the construction of the individual letters, their proportion being based on L.C. Evett's classic 1938 interpretation of the Trajan Column. They are presented in order of ease of construction, in four groups: the uprights, the obliques, the curves, and the loops.

The drawing instruments required are a T-square, an adjustable set square, and a set of compasses. Guidelines should be drawn lightly with hard graphite, with a softer graphite used for drawing curves. Once constructed, the letters can be brush-painted solid with India ink after outlining them with a technical or ruling pen.

N.B.: Inking mistakes can be retouched with white paint or typewriter correction fluid.

Apart from an upper guideline and a baseline, a halfway guideline should be used in the construction of such letters as F, H, Y, and K. However, an optical adjustment is made to the crossbar of the letter A and the middle arm of the letter E. Also, depth-of-loop lines for the letters R, P, and B deviate from the halfway line.

FHYK

AEP

2

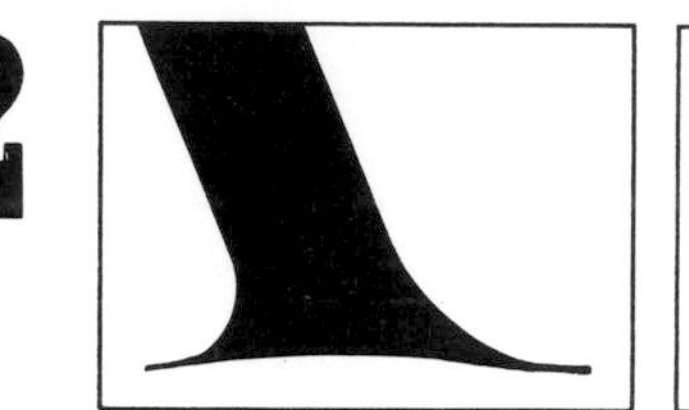

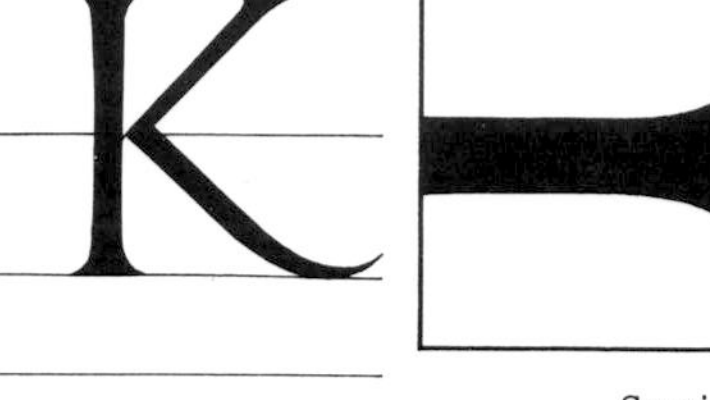

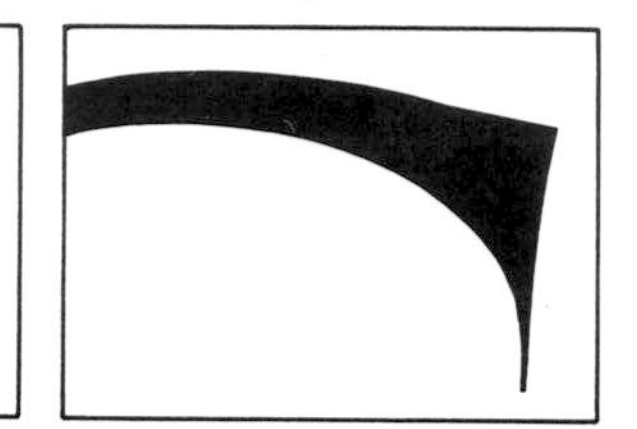

3

Serifs act to finish each stroke and also to help the eye pass along the line. There are several variations, but each should always extend the letter form, that is, appear to grow from the letter so that no hint of junction is evident.

4

QUI

Extended flourishes, or tails, occur on the letters Q, R, J, and K. Apart from that on the letter K, these swing below the baseline, with the extra-long version on the Q acting as a visual connection between it and U, its following letter.

The letter I is the easiest Roman letter form to construct. It is represented by a full-width upright stroke, its thickness being one-tenth of the chosen height of the letter size.

N.B.: All other stroke thicknesses are multiples of this width.

5

I

Two important optical effects that modify the appearance of the I and, indeed, all other full-stroke uprights in this alphabet, should be introduced. The first is the entasis, or slight concavity, of its sides, achieved by changing the drawing angle of the pencil against the raised edge of the ruler during construction. The second effect is a slight concavity of the face of each serif as it touches the upper and lower guidelines.

The Uprights: T, H

The length of the upper crossbar on the letter T is determined by the letter's height less two upright stroke thicknesses. The thickness of the crossbar is just over half that of the upright.

Bisect the crossbar to center the position of the upright.

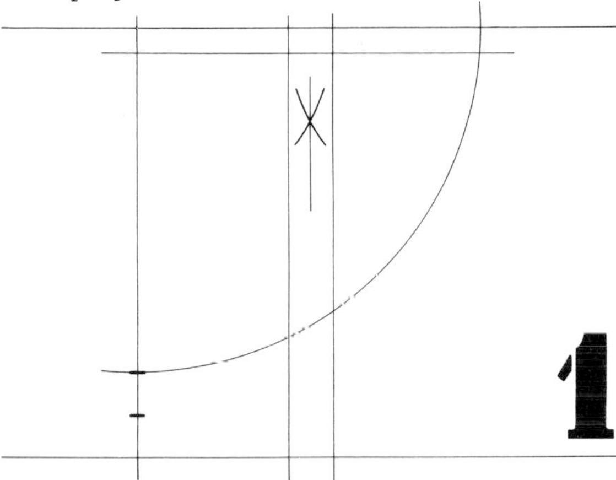

The result is an elegant and dynamic form, and one more easily spaced when forming words, by first locating the crossbar.

Notice that the crossbar is slanted slightly from left to right. This brings movement to the form and helps the eye read this letter in context with others. Having both a shortened and a slanted crossbar also avoids a potential top-heaviness commonly associated with this letter form.

The width of the letter H is determined by the letter's height less the thickness of one and a half upright strokes.

Although some designers prefer to draw the crossbar serifs so that they are symmetrical with the upright, a further refinement is to angle them along an eighty-degree slope to vertical. Yet another refinement is the subtle elongation of the lower, left-arm serif and the upper, right-arm serif.

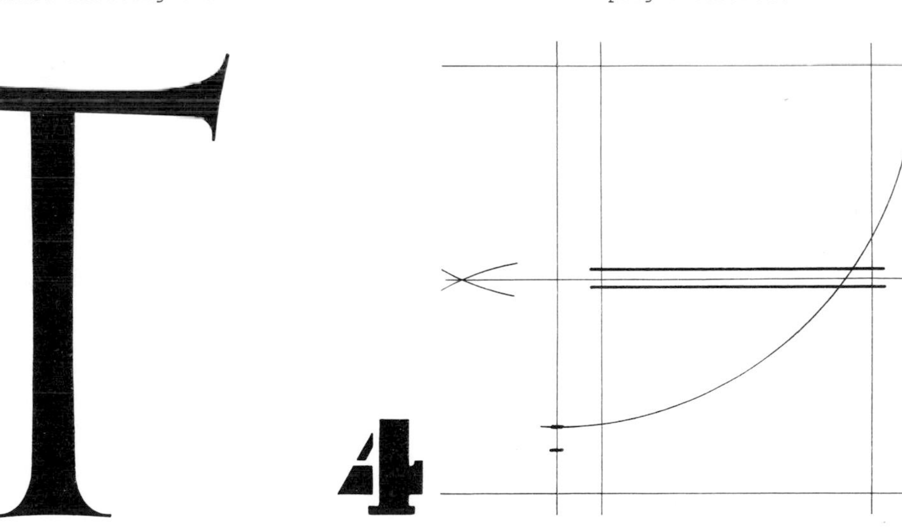

Essentially, the H is constructed like two letter I's connected by a central crossbar just over one-third the thickness of the upright.

The Uprights: L, E, F

To construct the letter L, first establish its full-width upright. One-half its height gives the extent of its arm, the thickness being roughly half that of the upright. This thickness is maintained for all the arms in the letters E and F.

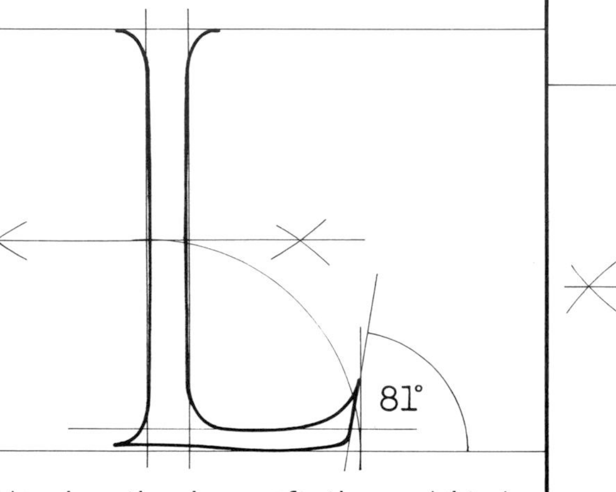

Notice how the base of the upright is raised with a subtle curve along the baseline, and also note the inclination of the serif's face.

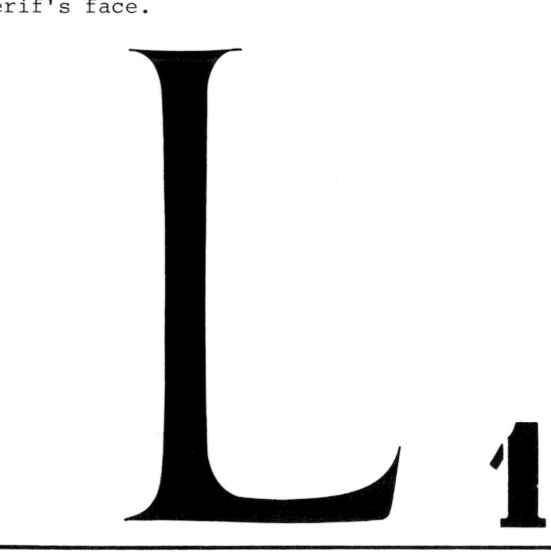

1

2

The letter E extends the construction of the letter L. Its middle arm is centered on a line bisecting the top guideline and a line formed one-half of a full-width stroke above the baseline. This dimension also gives the length of the base arm, the extent of the top arm being found by measuring back one-half of a full-width stroke.

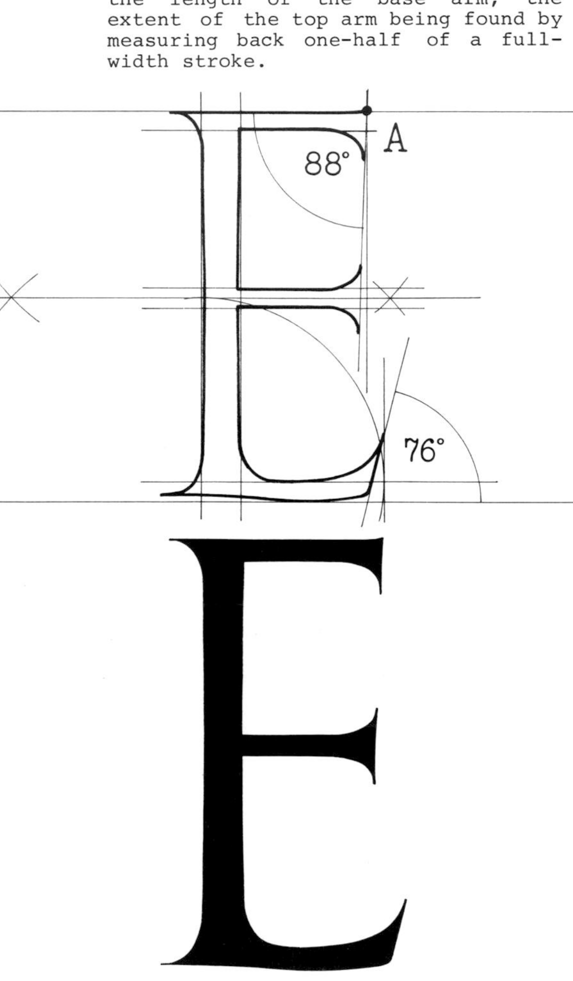

Note that the inclination of the serifs on the upper two arms is described by a line projected from A at 88 degrees to the horizontal.

The height of the letter F is bisected to give the center line for its lower arm. Point **B** is found by measuring one-half of a full-stroke's width up from the halfway line. The dimension from this point to the top guideline finds the extent of the upper arm.

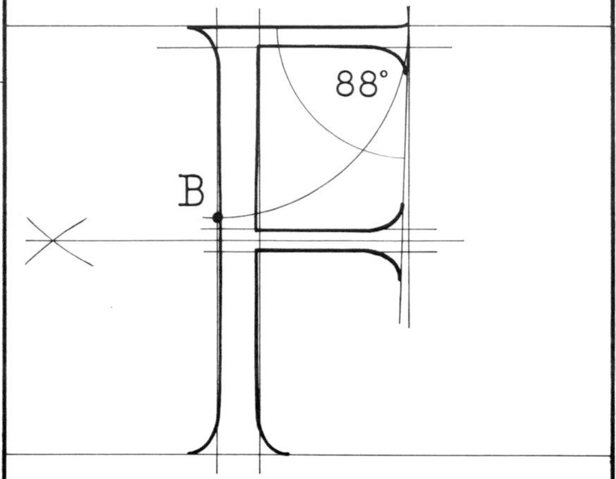

N.B.: The middle arm of the letter E is located higher than its counterpart on the letter F. However, in each case, their extent and the angle of their serifs is found by a line angled at 88 degrees.

3

The Obliques: K, Y

After establishing the full-width upright of the letter K, project its arms from a point on the halfway line. The upper arm is drawn at an angle of 46 degrees to the horizontal, the lower arm at an angle of 44 degrees.

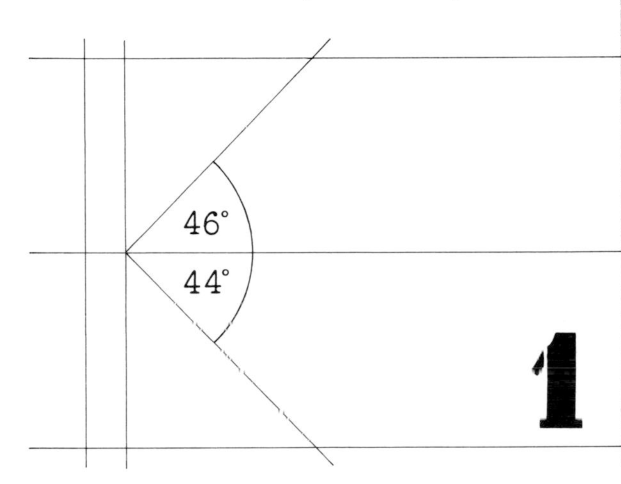

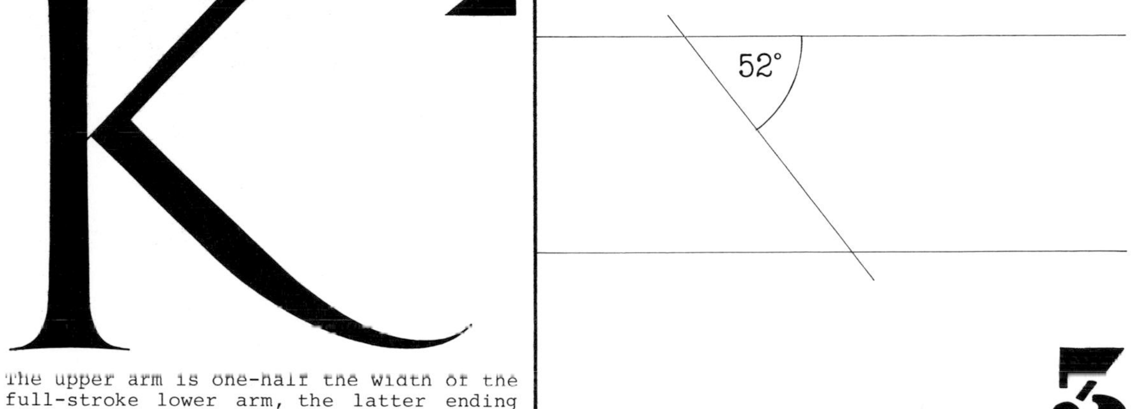

The upper arm is one-half the width of the full-stroke lower arm, the latter ending with a flourish similar to that in the letter R.

To construct the letter Y, first establish the outer edge of its left arm. This is described by a line drawn at 52 degrees from a point on the top guideline down to the halfway line.

From the point of intersection with the halfway line, measure one full stroke's width to the right. From this point the outside edge of the right arm is projected at an angle of 54 degrees. Add lines to describe the inside edges of the left and right arms. The arms are full- and half-stroke width, respectively.

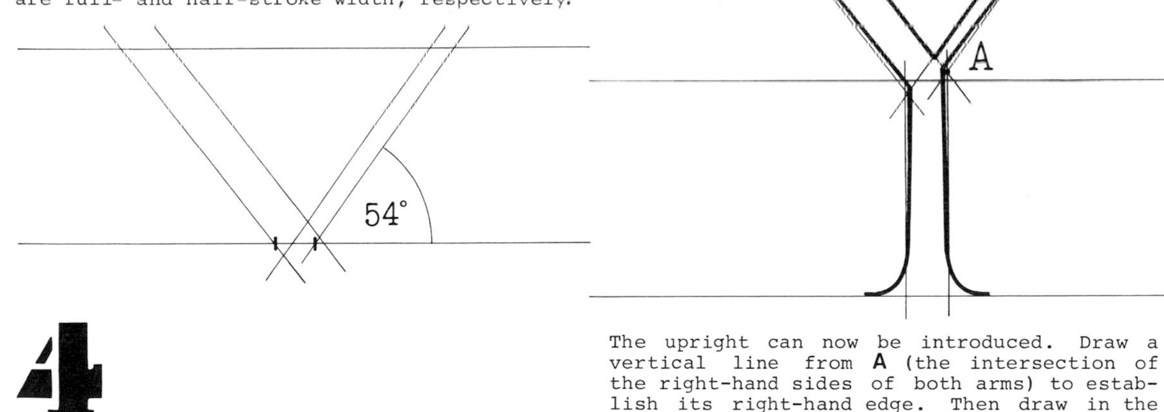

The upright can now be introduced. Draw a vertical line from **A** (the intersection of the right-hand sides of both arms) to establish its right-hand edge. Then draw in the left edge of the full-width stroke.

Slight entasis occurs on the upper stem of the upright stroke and the outer side of the lower arms.

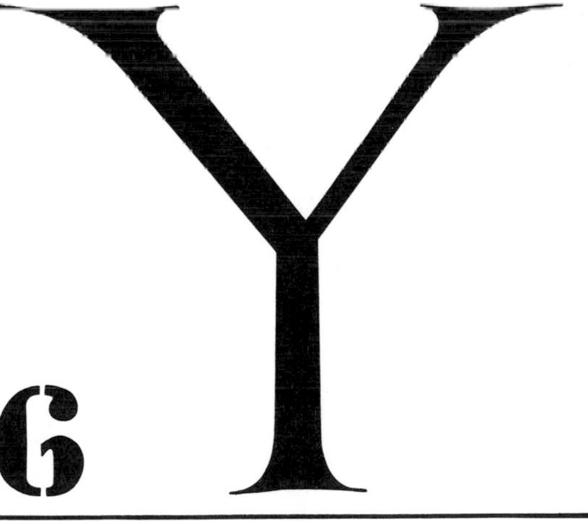

The Obliques: N, X

The apex of the letter N occurs at a point on a line drawn one-quarter of a wide stroke's width above the upper guideline.

From this point, a diagonal is then drawn at an angle of 46 degrees. One full-stroke width below this, a second diagonal is drawn to connect with a line drawn one-half of a full stroke's width under the baseline. This point of intersection finds the outer edge of the right upright.

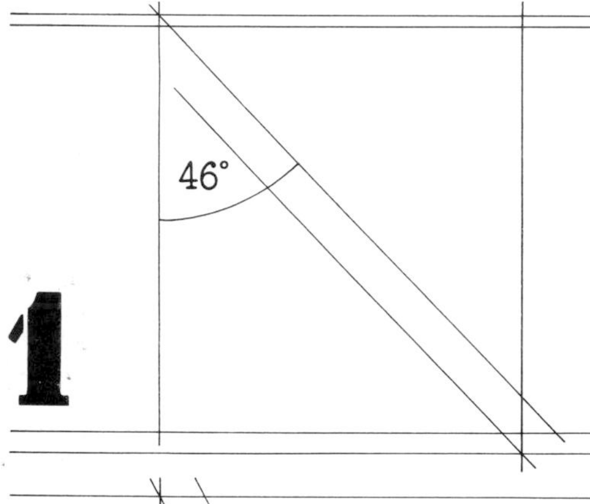

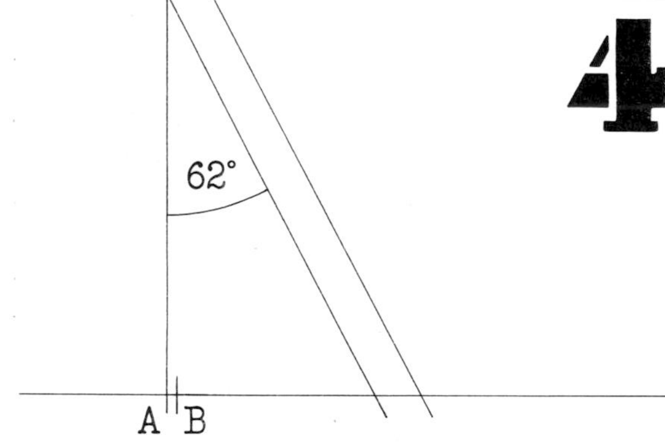

To construct the letter X, first mark off two points one-quarter of a wide stroke apart on the baseline. From point **A**, project a vertical line to the top guideline. From this point, project a right-to-left diagonal at 62 degrees. This describes the left edge of a full-width stroke.

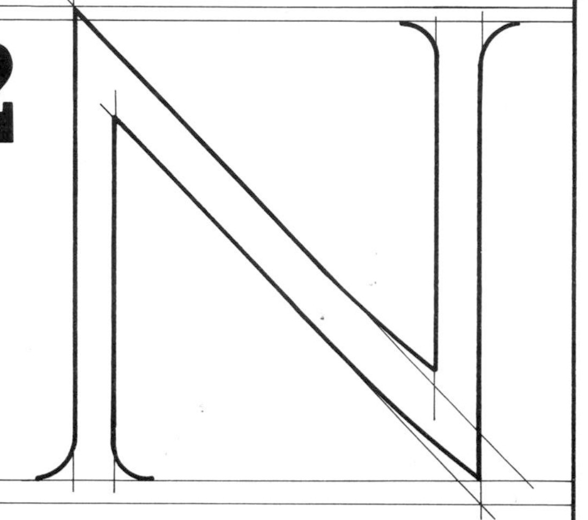

When the diagonal stroke is drawn, a subtle curve is introduced along its length to bring its lower point onto the baseline.

From point **B** on the baseline, project a left-to-right diagonal at an angle of 64 degrees. This describes the left edge of a stroke one-half the full-stroke width.

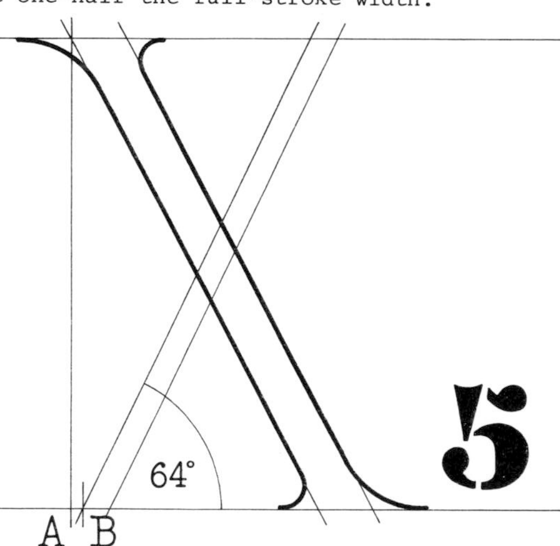

Both the upright strokes are seven-eighths of a wide stroke in thickness.

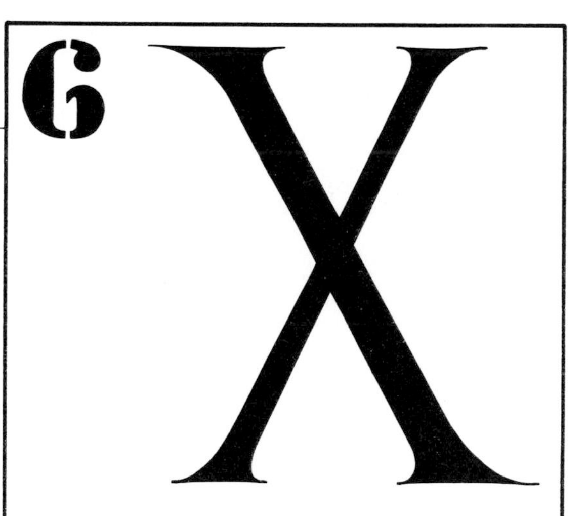

As with other oblique serifs in the Roman alphabet, the outer versions are slightly larger and extended.

The Obliques: A, V

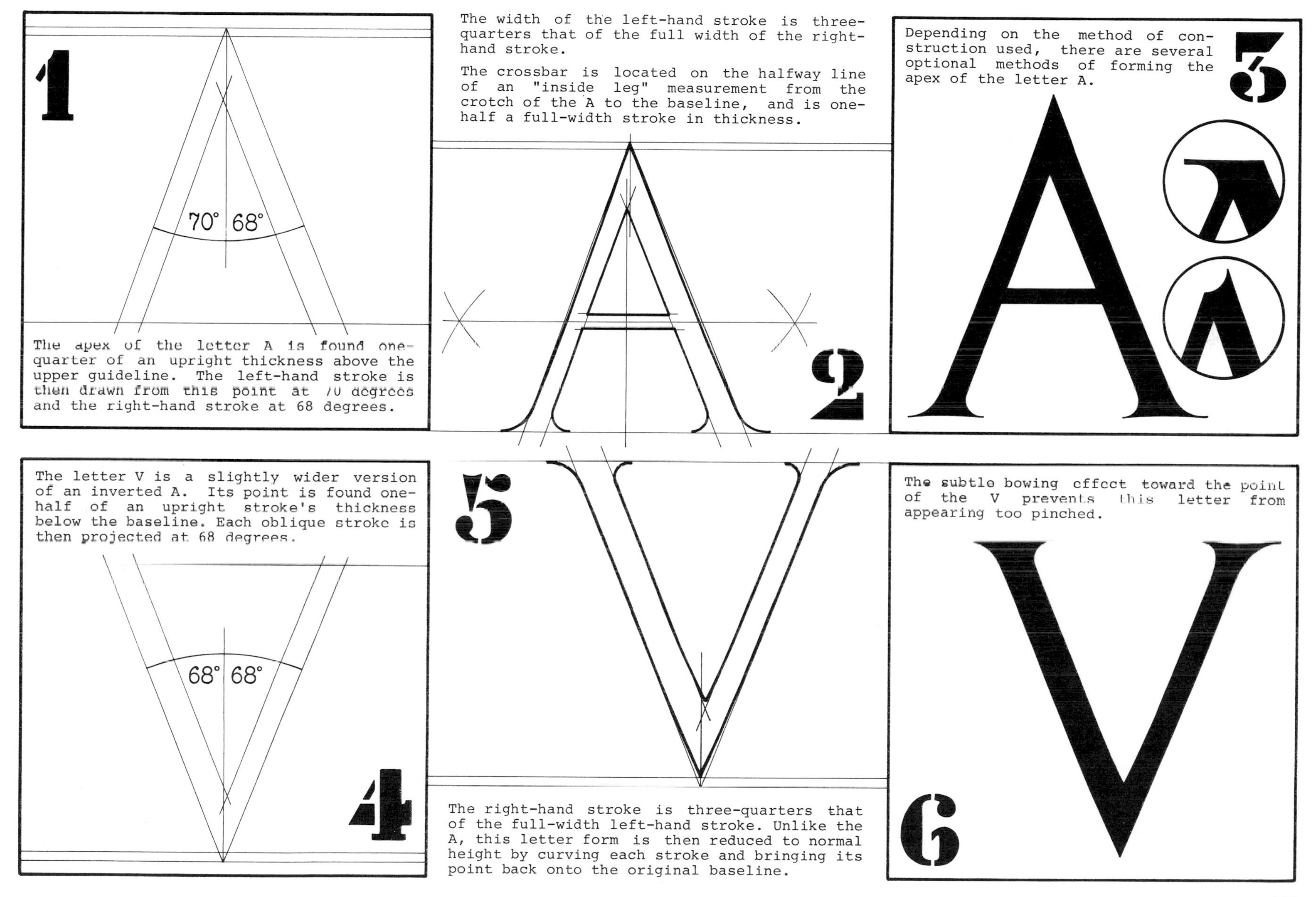

The apex of the letter A is found one-quarter of an upright thickness above the upper guideline. The left-hand stroke is then drawn from this point at 70 degrees and the right-hand stroke at 68 degrees.

The width of the left-hand stroke is three-quarters that of the full width of the right-hand stroke.

The crossbar is located on the halfway line of an "inside leg" measurement from the crotch of the A to the baseline, and is one-half a full-width stroke in thickness.

Depending on the method of construction used, there are several optional methods of forming the apex of the letter A.

The letter V is a slightly wider version of an inverted A. Its point is found one-half of an upright stroke's thickness below the baseline. Each oblique stroke is then projected at 68 degrees.

The right-hand stroke is three-quarters that of the full-width left-hand stroke. Unlike the A, this letter form is then reduced to normal height by curving each stroke and bringing its point back onto the original baseline.

The subtle bowing effect toward the point of the V prevents this letter from appearing too pinched.

The Obliques: W, Z

1 The construction of the letter W is, essentially, that of two overlapping V's. First, construct a letter V as described on page 17.

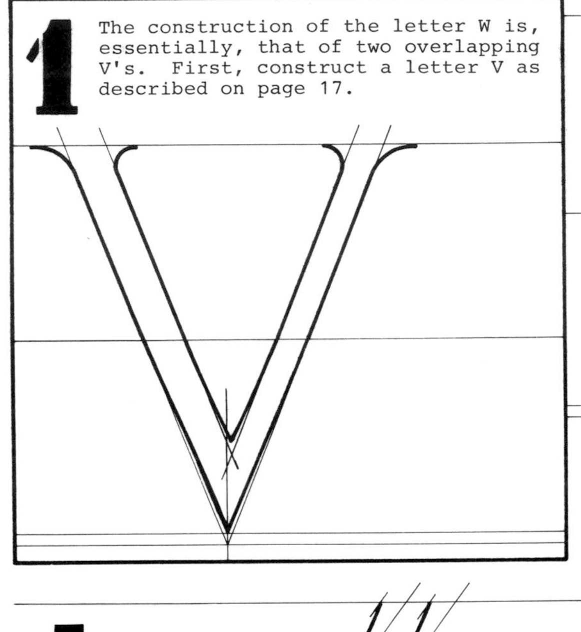

2 Then construct a second letter V so that the left edge of its left arm cuts the right edge of the right arm of the first V on the halfway line.

3 Notice how the points of the two V's that comprise this letter are bowed to bring them onto the original baseline.

4

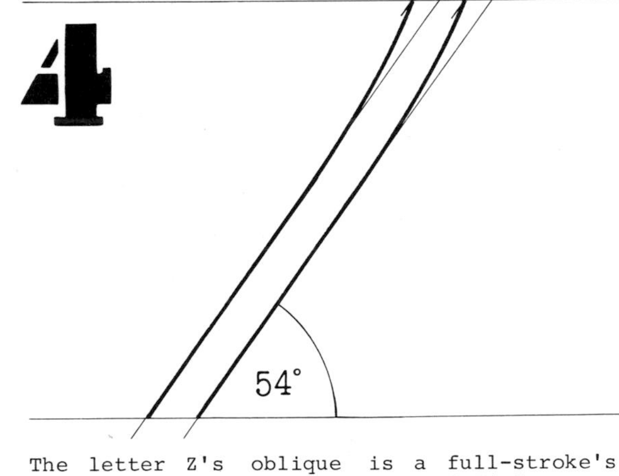

The letter Z's oblique is a full-stroke's width and is drawn at an angle of 54 degrees. From the points where each line connects with the upper guideline, measure back one-half a full-stroke's width. Both oblique lines are then curved along their length to meet these new points.

5 The points where the left side of the oblique stroke meets the baseline and the (now curved) right side of the stroke meets the top line find the extent of the Z's horizontal bars. Both bars are one-half the thickness of a wide stroke.

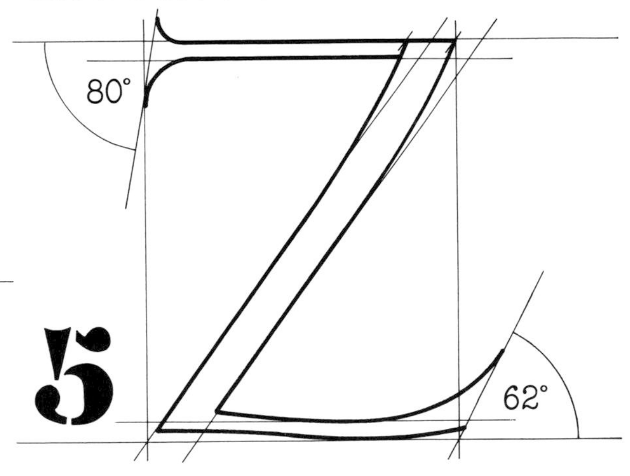

6 Notice how the base of the oblique stroke sits slightly above the baseline and the lower bar rocks gracefully on the baseline. Also, note the subtle tilting of both serif faces, the lower being more acute and subtly elongated at its upper point.

The Obliques: M

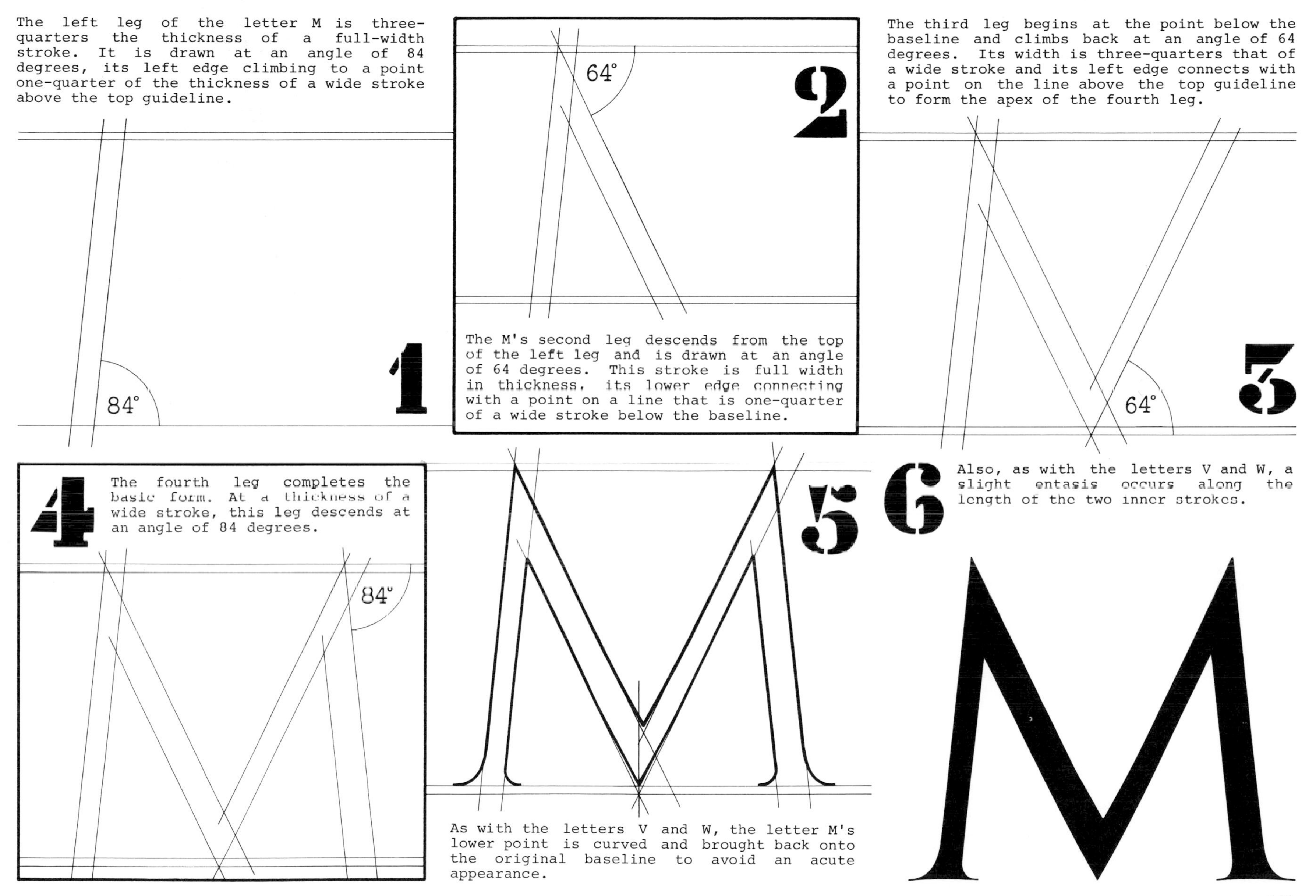

The left leg of the letter M is three-quarters the thickness of a full-width stroke. It is drawn at an angle of 84 degrees, its left edge climbing to a point one-quarter of the thickness of a wide stroke above the top guideline.

The M's second leg descends from the top of the left leg and is drawn at an angle of 64 degrees. This stroke is full width in thickness, its lower edge connecting with a point on a line that is one-quarter of a wide stroke below the baseline.

The third leg begins at the point below the baseline and climbs back at an angle of 64 degrees. Its width is three-quarters that of a wide stroke and its left edge connects with a point on the line above the top guideline to form the apex of the fourth leg.

The fourth leg completes the basic form. At a thickness of a wide stroke, this leg descends at an angle of 84 degrees.

As with the letters V and W, the letter M's lower point is curved and brought back onto the original baseline to avoid an acute appearance.

Also, as with the letters V and W, a slight entasis occurs along the length of the two inner strokes.

The Curves: O

The construction of the letter O begins on a center point on the halfway line around which a full letter-height circle is described.

Through its center, draw an inclined vertical with an angle of 82 degrees. A circle with a radius of one full stroke's width is then described around the center giving points **A**, **B**, **C**, and **D**.

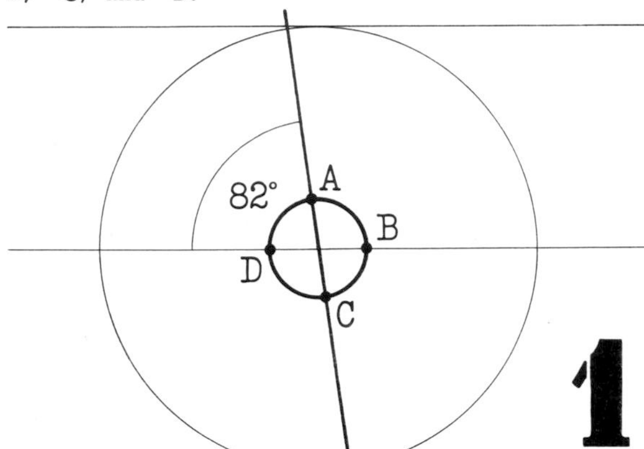

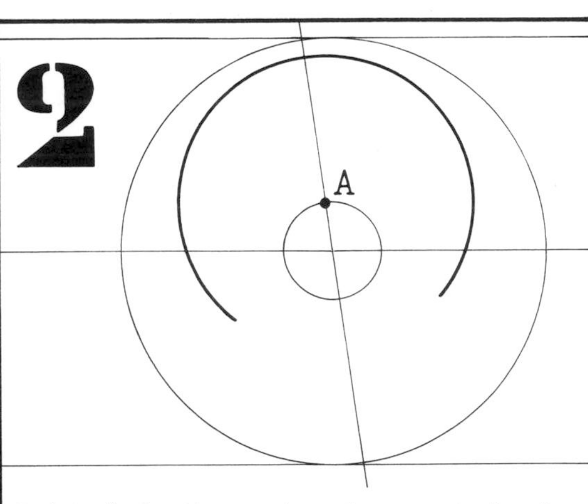

Point **A** is the center for a circle describing the inside of the upper part of the letter. Its radius is found by measuring three-eighths of a full stroke's width from the upper guideline.

Using the same radius, transfer the compass point to point **C** to describe the inside of the lower part of the letter.

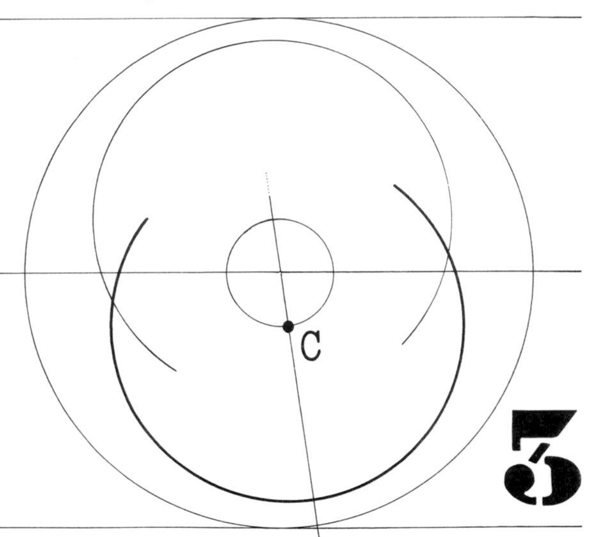

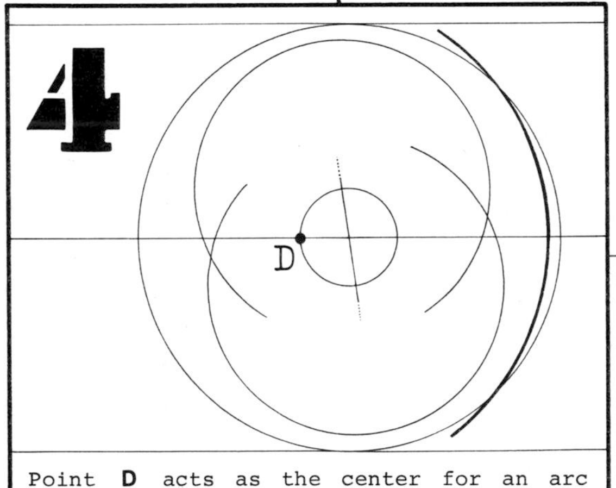

Point **D** acts as the center for an arc that reduces the right-hand side of the letter. Its radius is found by measuring back one-quarter of a full stroke's width from the intersection of the full circle on the halfway line.

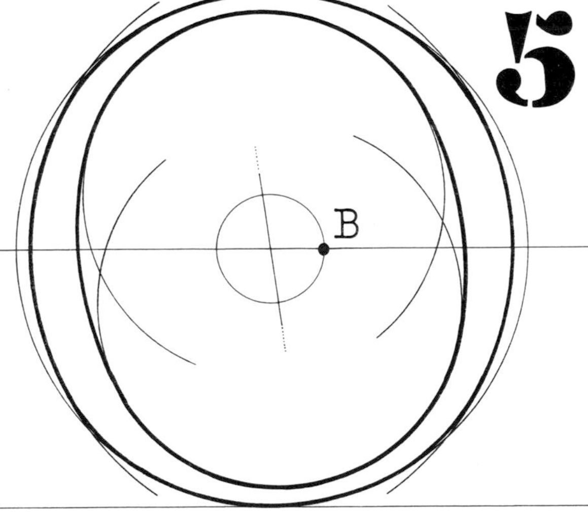

Using the same radius, transfer the compass point to point **B** and describe the same arc on the left-hand side of the letter.

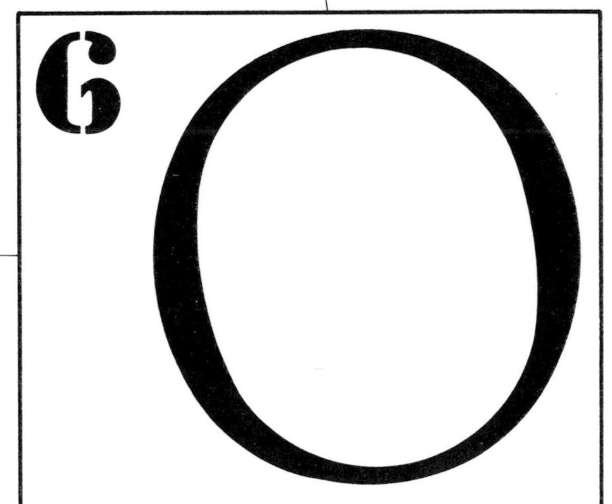

The thinner parts of the letter O are a little less than one-half a full stroke's width. These should enlarge gradually into the full thickness.

The Curves: Q, D

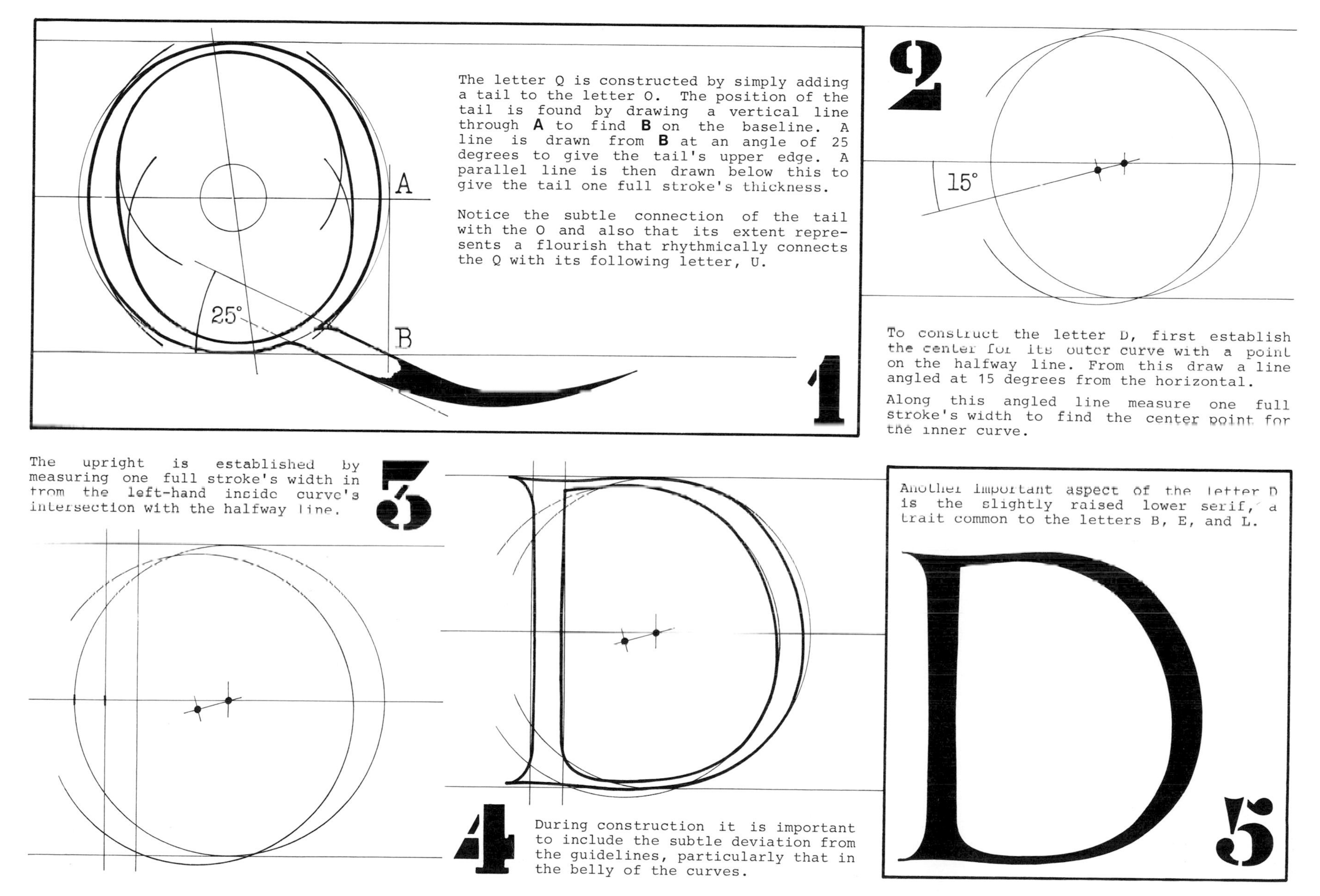

The letter Q is constructed by simply adding a tail to the letter O. The position of the tail is found by drawing a vertical line through **A** to find **B** on the baseline. A line is drawn from **B** at an angle of 25 degrees to give the tail's upper edge. A parallel line is then drawn below this to give the tail one full stroke's thickness.

Notice the subtle connection of the tail with the O and also that its extent represents a flourish that rhythmically connects the Q with its following letter, U.

To construct the letter D, first establish the center for its outer curve with a point on the halfway line. From this draw a line angled at 15 degrees from the horizontal.

Along this angled line measure one full stroke's width to find the center point for the inner curve.

The upright is established by measuring one full stroke's width in from the left-hand inside curve's intersection with the halfway line.

During construction it is important to include the subtle deviation from the guidelines, particularly that in the belly of the curves.

Another important aspect of the letter D is the slightly raised lower serif, a trait common to the letters B, E, and L.

The Curves: C, G

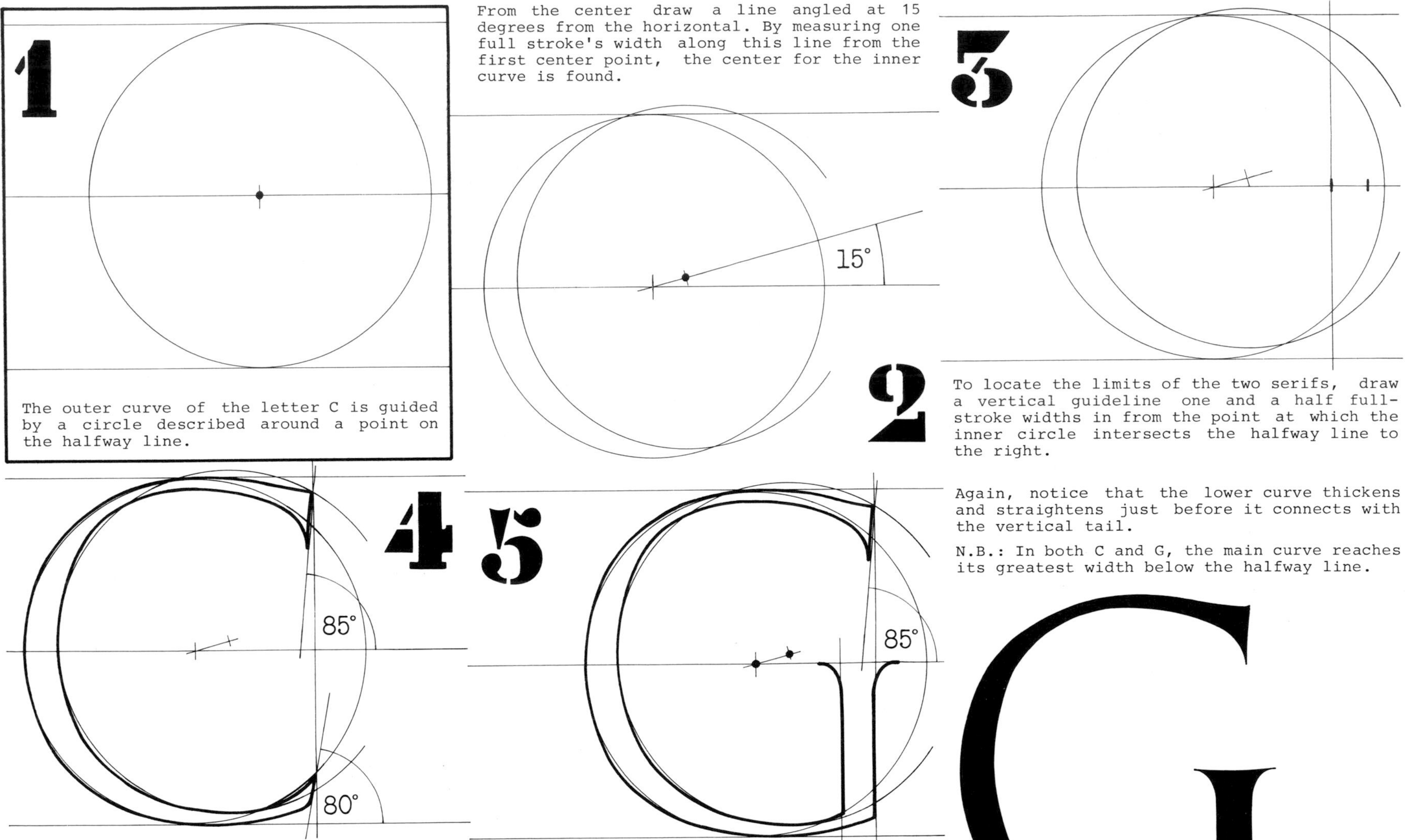

The outer curve of the letter C is guided by a circle described around a point on the halfway line.

From the center draw a line angled at 15 degrees from the horizontal. By measuring one full stroke's width along this line from the first center point, the center for the inner curve is found.

To locate the limits of the two serifs, draw a vertical guideline one and a half full-stroke widths in from the point at which the inner circle intersects the halfway line to the right.

Make sure that the upper and lower curves straighten and thicken slightly as they merge into the serifs.

N.B.: The face of each serif is angled slightly from the vertical, and the upper arm is heavier than the lower one.

The construction of the letter G is identical with that of the C up until step 3. The difference is the G's vertical tail, which is one full stroke in width. Its outer edge coincides with the vertical guideline.

Again, notice that the lower curve thickens and straightens just before it connects with the vertical tail.

N.B.: In both C and G, the main curve reaches its greatest width below the halfway line.

The Curves and Loops: J, U

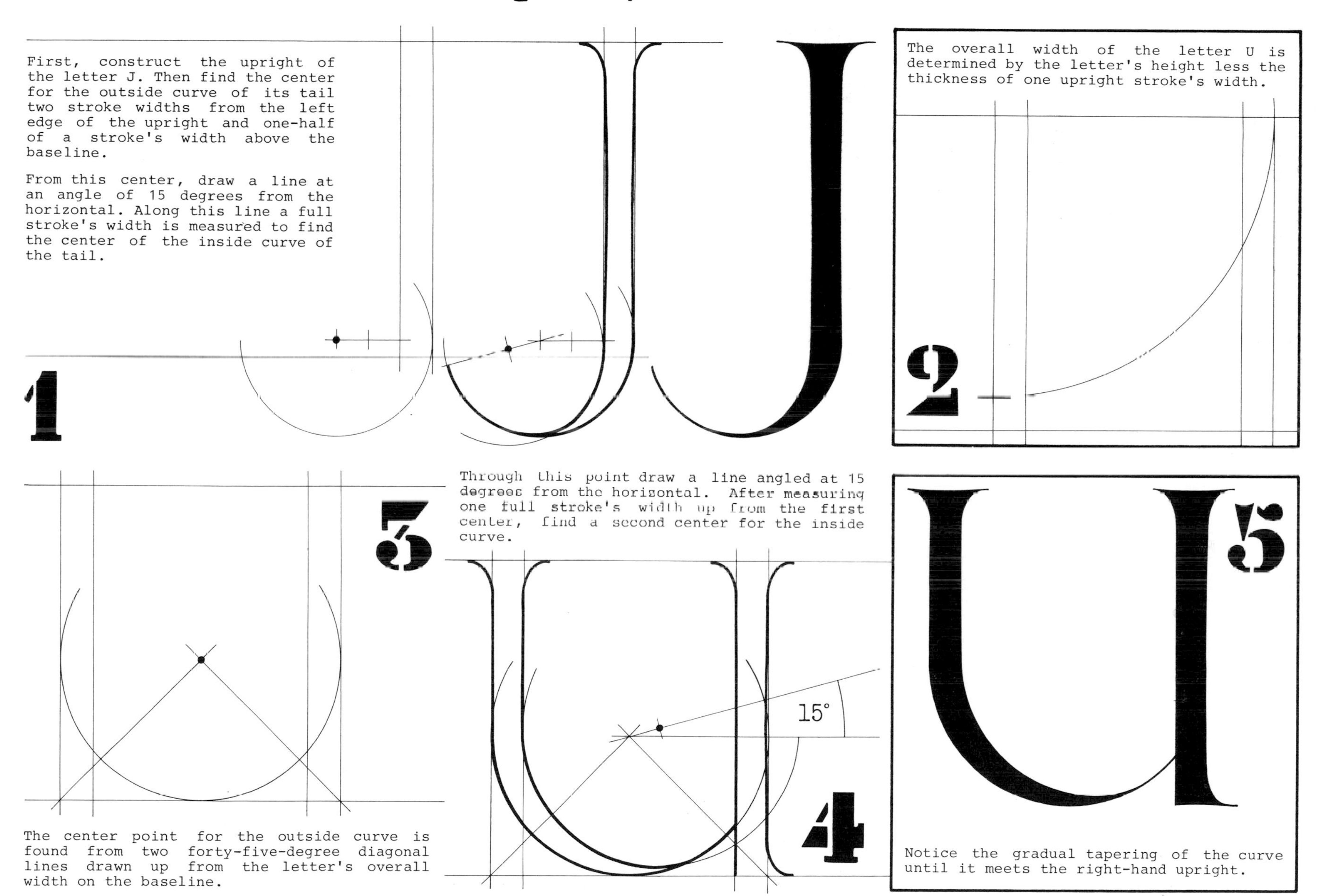

First, construct the upright of the letter J. Then find the center for the outside curve of its tail two stroke widths from the left edge of the upright and one-half of a stroke's width above the baseline.

From this center, draw a line at an angle of 15 degrees from the horizontal. Along this line a full stroke's width is measured to find the center of the inside curve of the tail.

The overall width of the letter U is determined by the letter's height less the thickness of one upright stroke's width.

The center point for the outside curve is found from two forty-five-degree diagonal lines drawn up from the letter's overall width on the baseline.

Through this point draw a line angled at 15 degrees from the horizontal. After measuring one full stroke's width up from the first center, find a second center for the inside curve.

Notice the gradual tapering of the curve until it meets the right-hand upright.

The Curves and Loops: S, P

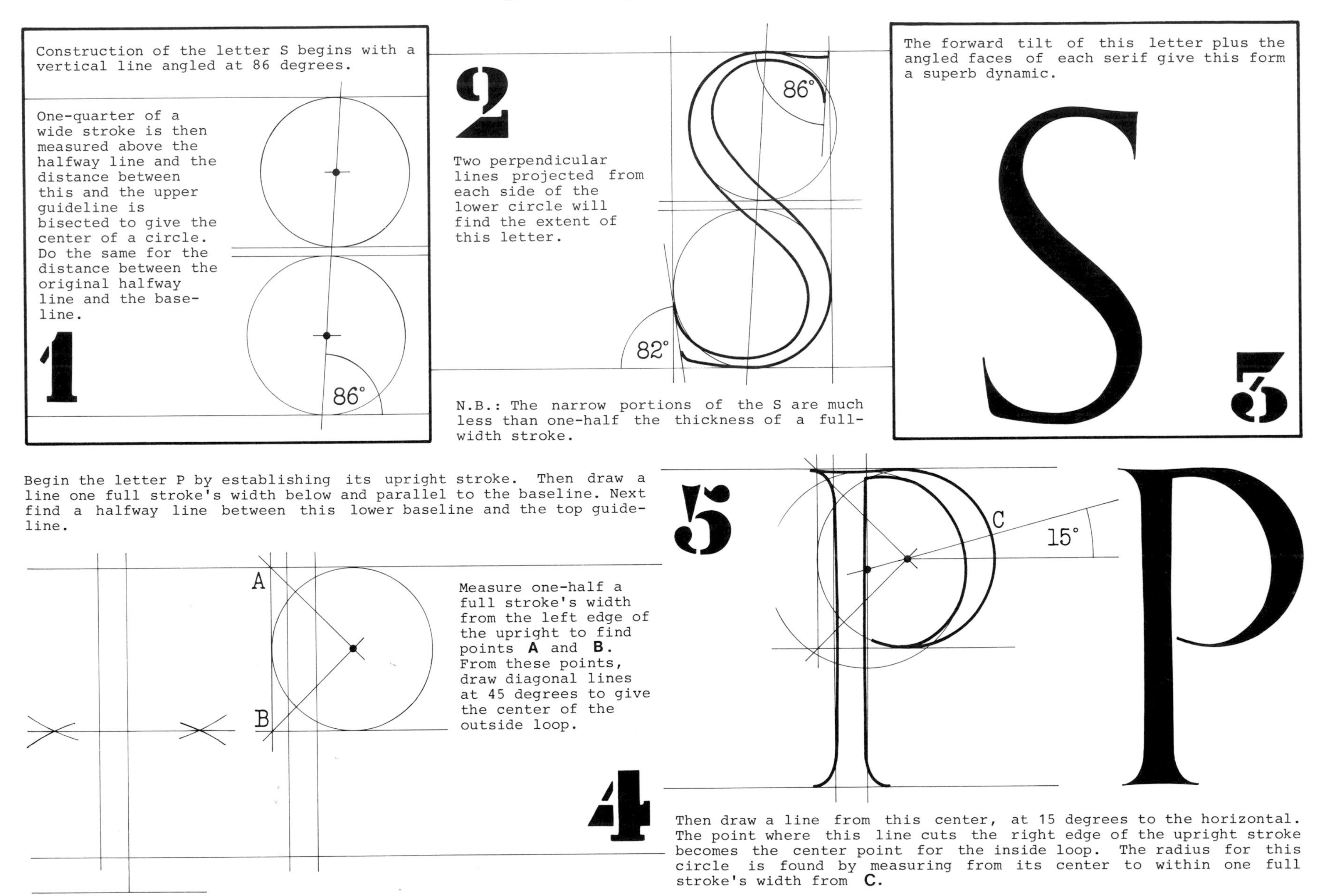

Construction of the letter S begins with a vertical line angled at 86 degrees.

One-quarter of a wide stroke is then measured above the halfway line and the distance between this and the upper guideline is bisected to give the center of a circle. Do the same for the distance between the original halfway line and the baseline.

Two perpendicular lines projected from each side of the lower circle will find the extent of this letter.

N.B.: The narrow portions of the S are much less than one-half the thickness of a full-width stroke.

The forward tilt of this letter plus the angled faces of each serif give this form a superb dynamic.

Begin the letter P by establishing its upright stroke. Then draw a line one full stroke's width below and parallel to the baseline. Next find a halfway line between this lower baseline and the top guideline.

Measure one-half a full stroke's width from the left edge of the upright to find points **A** and **B**. From these points, draw diagonal lines at 45 degrees to give the center of the outside loop.

Then draw a line from this center, at 15 degrees to the horizontal. The point where this line cuts the right edge of the upright stroke becomes the center point for the inside loop. The radius for this circle is found by measuring from its center to within one full stroke's width from **C**.

The Loops: B

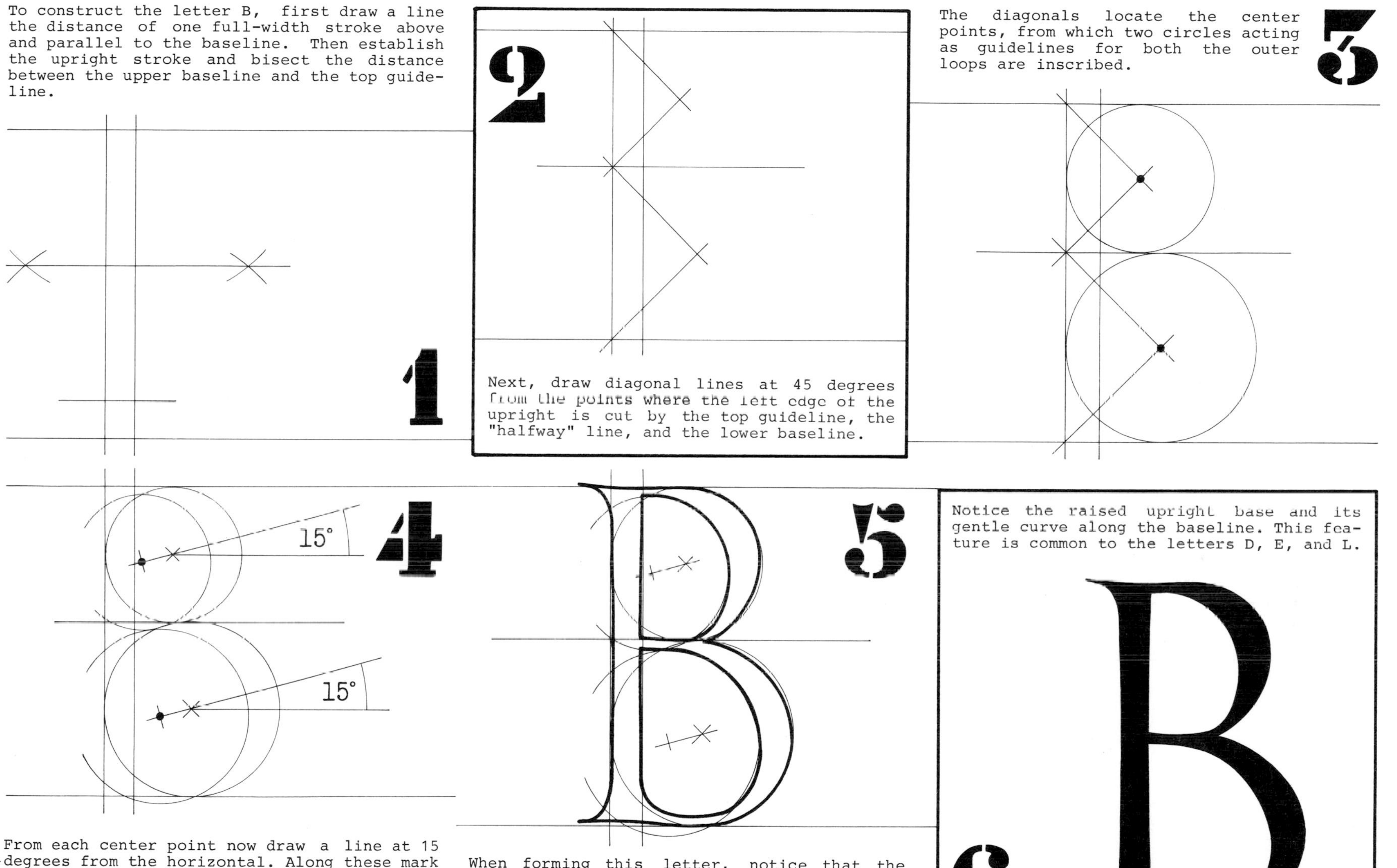

To construct the letter B, first draw a line the distance of one full-width stroke above and parallel to the baseline. Then establish the upright stroke and bisect the distance between the upper baseline and the top guideline.

Next, draw diagonal lines at 45 degrees from the points where the left edge of the upright is cut by the top guideline, the "halfway" line, and the lower baseline.

The diagonals locate the center points, from which two circles acting as guidelines for both the outer loops are inscribed.

From each center point now draw a line at 15 degrees from the horizontal. Along these mark two further points one full stroke's width to the left of each original center. These give the centers for circles acting as guidelines for both the inner loops.

When forming this letter, notice that the horizontal connecting bars have individual qualities that extend the nature of the loops. The thickness of the connections is about a third that of the full stroke's width.

Notice the raised upright base and its gentle curve along the baseline. This feature is common to the letters D, E, and L.

The Loops: R

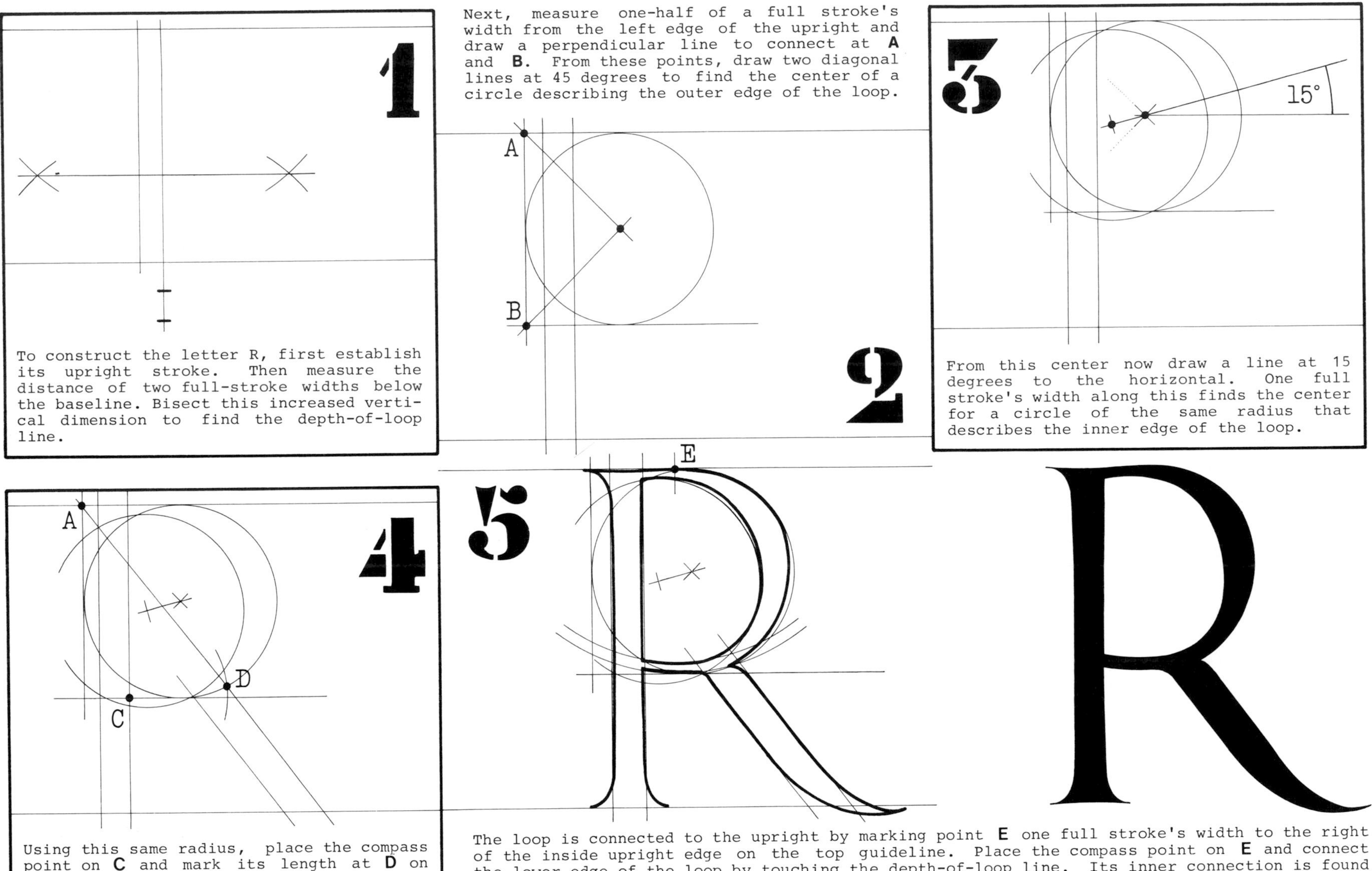

To construct the letter R, first establish its upright stroke. Then measure the distance of two full-stroke widths below the baseline. Bisect this increased vertical dimension to find the depth-of-loop line.

Next, measure one-half of a full stroke's width from the left edge of the upright and draw a perpendicular line to connect at **A** and **B**. From these points, draw two diagonal lines at 45 degrees to find the center of a circle describing the outer edge of the loop.

From this center now draw a line at 15 degrees to the horizontal. One full stroke's width along this finds the center for a circle of the same radius that describes the inner edge of the loop.

Using this same radius, place the compass point on **C** and mark its length at **D** on the outer edge of the loop. A line drawn from **A** to cut **D** gives the outer edge of the tail.

The loop is connected to the upright by marking point **E** one full stroke's width to the right of the inside upright edge on the top guideline. Place the compass point on **E** and connect the lower edge of the loop by touching the depth-of-loop line. Its inner connection is found one-third of a full stroke's width away.

N.B.: There is a subtle curve at the point of connection between tail and loop. The tail is a full stroke's width and its flourish ends below the baseline.

Lowercase Roman and Sans Serif Lettering

This lowercase, or minuscule, alphabet accompanies the Roman capitals described on pages 12-26. Essentially, lowercase characters retain the proportional principles that govern the construction of the capitals. However, the half serifs on the letters b, d, h, and so on are slightly angled from the horizontal. Also, the full-length vertical strokes are usually drawn to a height just above that of the capitals.

N.B.: As opposed to Roman capitals, which enjoy more generous spacing, the lowercase letters are easier to group, because they contain a less varied range of forms. Their mainly rounded shapes, and the spaces they define, appear rather like links in a chain: too compacted and they merge, too loose and the chain breaks.

1

abcdefghijklmnopqrst
uvwxyz 1234567890

2 abcdefghijklmnopqrst
uvwxyz 1234567890

1234567890

In this lowercase sans serif alphabet, the full-length vertical strokes remain at the same overall height as their capital counterparts. Also, numerals used in sans serif alphabets are usually the same height as capital letters. These examples, however, display both the regular and the more useful nonranging numerals. In both Roman and sans serif styles of nonranging numerals, a lowercase O is used for zero. Also, remember that the tails, or descenders, on lowercase letters (such as the j, p, and q) dictate a wider line spacing than usual in order to avoid collision with the ascenders.

Hints when Spacing Lettering

1

To make a word appear complete and, when combined with others, as a unit in a design, it is essential that the areas of space between its letters be balanced. However, this is impossible to achieve when letters are equally spaced.

Letter and word spacing is an exercise in optical illusion, the eye reading the gaps between letters and words as space and not as distance. The control of spacing therefore should be governed by the eye and not by the ruler.

UTAH STATE

4

Beware of the potential spatial "hiccup" that can occur when vertical letter forms are followed by oblique or rounded letters.

ATTAR
LILY

N.B.: When such letters as TA and LY come together in a tight formation, they can be allowed to overlap slightly.

2

MINIMUM ✗

MINIMUM ✓

Spacing problems lurk where certain letter forms are brought together; for example, when closely related sequences of upright strokes occur. In such events more space should be allowed, to retain an even pattern.

AVOCADO ✗
AVOCADO ✓
INFINITE ✗
INFINITE ✓
CITRIC ✗
CITRIC ✓

5

When such letters as AV, IN, and CI come together, they produce minimal surrounding space. The potential uneven word pattern is avoided by spacing out the letters.

3

Conversely, when letters that include large areas of space as part of their construction come together, such as two T's, the other letters in the word should be spatially eased in compensation.

LETTER ✗

LETTER ✓

6

The golden rule is to preplan the word or word sequence prior to construction so that erratic patterns that might distract the eye can be anticipated and avoided.

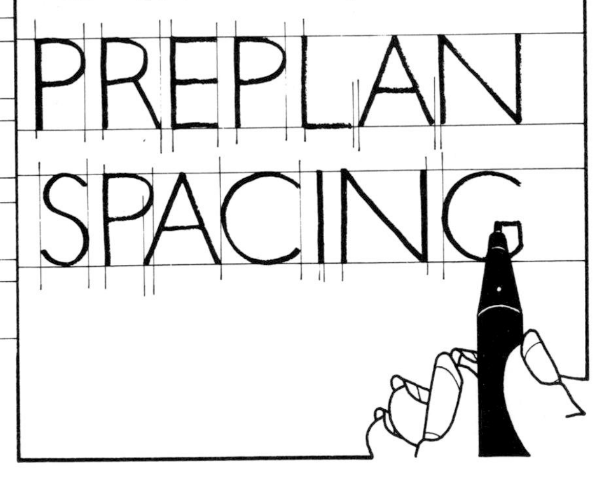

Hints when Designing Lettering Layouts

Decisions concerning the size of format and lettering together with lettering style and color are related directly to the intended function and the predicted viewing distance of the message in question.

A point of departure in the design of, for instance, a poster is to first group its constituent words into degrees of meaning and levels of importance. These can then be translated into an appropriate use of capitals and lowercase, and by lettering style, weight of letters, and so on.

Designers should be aware of an optical effect that can cause units of lettering (and, indeed, complete layouts) to disengage visually from their backgrounds. This is an aspect of the figure-ground illusion that can occur when words or shapes encroach on or come in contact with the edge of the format.

2

Using words or groups of words as design units, a trial layout can now be begun. This stage can be accelerated using a mockup of scaled strips cut from dark paper to simulate blocks of lettering. These are maneuvered on the format until a satisfactory layout is achieved. Several layouts should be tried, the main aim always being that of effective simplicity.

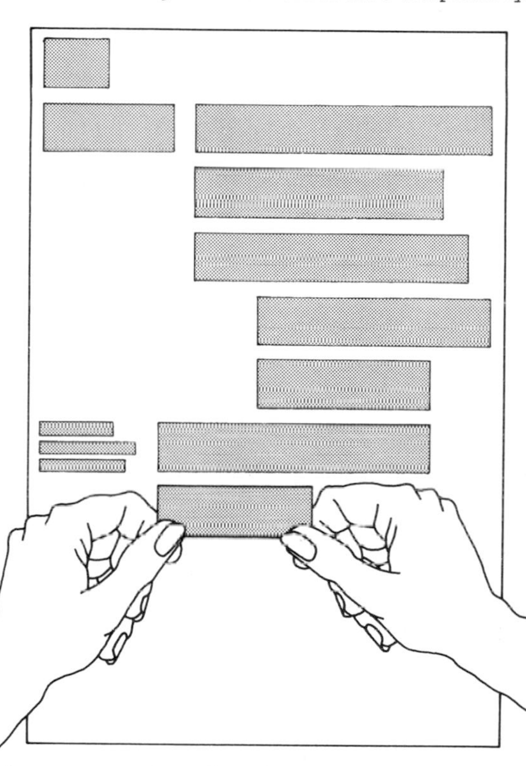

As a general rule, layouts should avoid the boredom of symmetry, the illegibility of vertical lettering, the visual "holes" caused by large areas of empty space or, conversely, heavy concentrations of lettering, as well as the potential confusion of jumbled layouts.

The spacing between lines of lettering is as important a design consideration as that between letters and words, because erratic line spacing can disrupt the continuity of meaning in the message. Line spacing is best designed in conjunction with the size of lettering in use, line spacing contracting to connect lines of meaning and expanding to separate units of information.

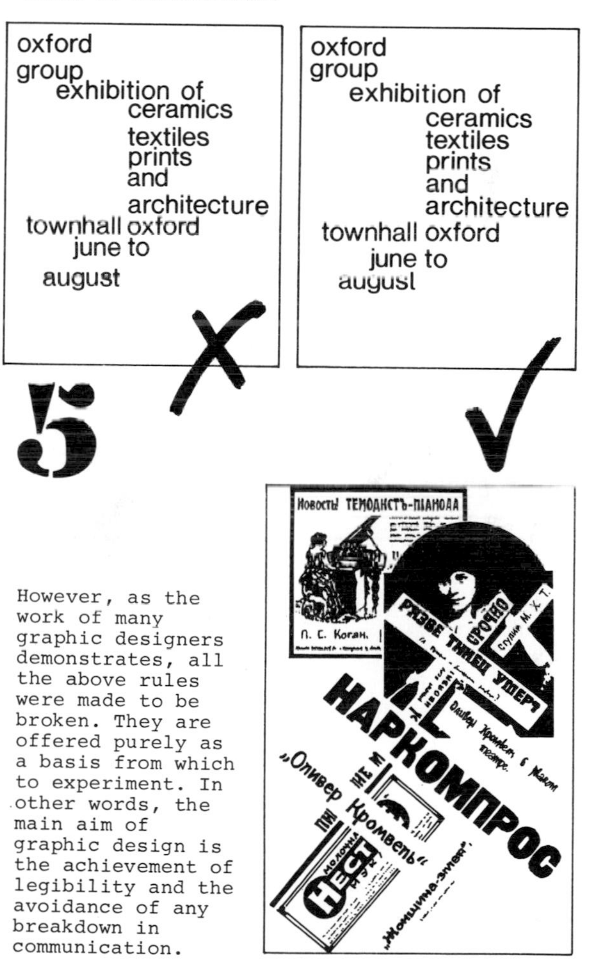

However, as the work of many graphic designers demonstrates, all the above rules were made to be broken. They are offered purely as a basis from which to experiment. In other words, the main aim of graphic design is the achievement of legibility and the avoidance of any breakdown in communication.

Lettering as a Graphic Element

1 As the modern environment bombards our retinas with myriad forms of visual communication, and as much of this barrage is comprised of lettering, the considered selection and incorporation of lettering in graphic design is crucial. The letter form is a prefabricated element whose source of alphabet styles is as limitless as is the number of design permutations to which it may be subjected. The designer's task, therefore, is to organize lettering, either with or without accompanying images, in a manner that is clear, legible, and visually compelling.

2 Lettering can, by itself, become a powerfully descriptive graphic element. The double function of lettering as both printed word and pictorial image can result from selecting letter forms sympathetic to the message, and the relationships of their spacing in word formation.

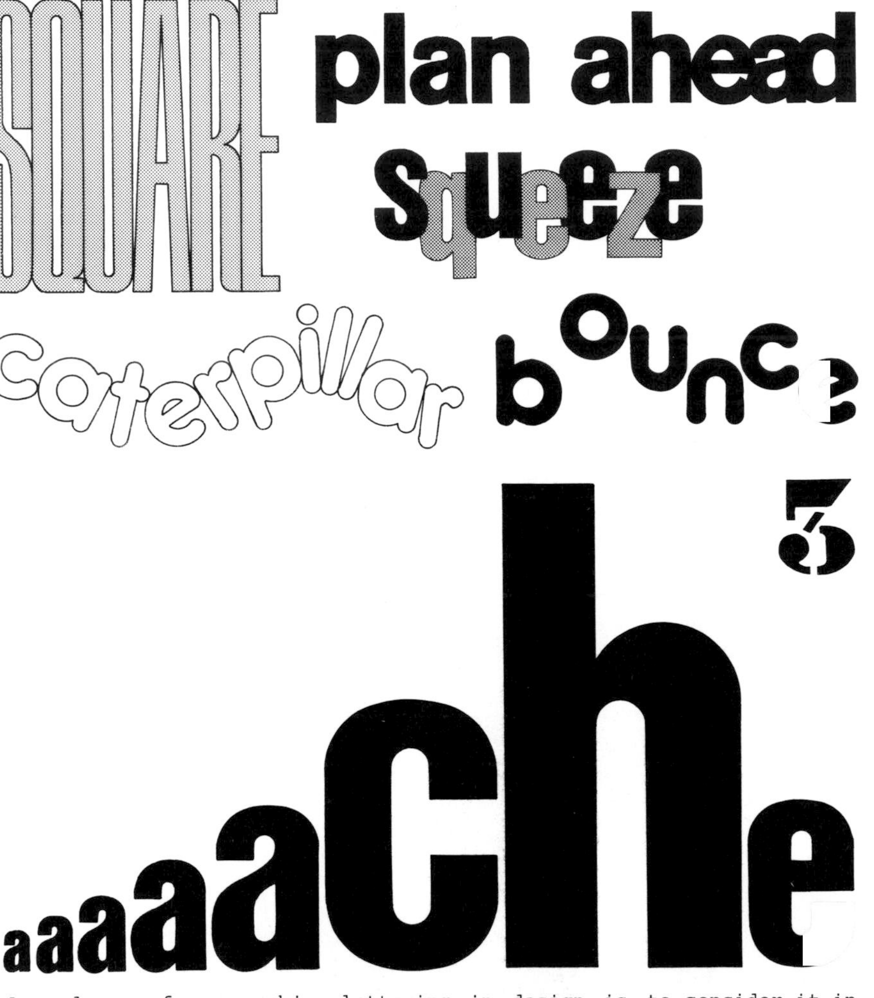

3 A good way of approaching lettering in design is to consider it in this dual role of symbol and of form. In this manner, letter forms become building blocks in the structuring of graphic compositions.

Lettering as a Graphic Element

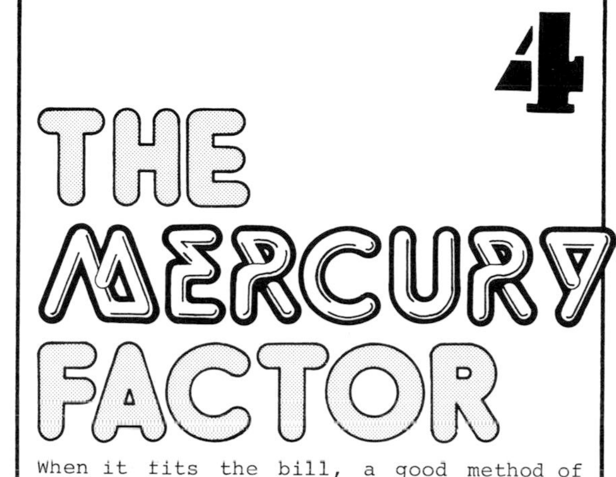

When it fits the bill, a good method of extending the symbol-form duality of lettering in posters and ads is to invest a key adjective with a descriptive treatment that is intrinsic to its meaning.

The apparent permanence of precise lettering can be successfully contrasted in size, color, and texture, with the spontancity of handwritten words, the transience of stenciled lettering, or the immediacy of the rubber stamp.

The latter can be introduced by cutting the word in reverse in a lino block and, after inking up, printing the word on the layout (see page 68).

edna bucket
the banana skin
a study of
the incurably
disaster-prone

5

Another method of drawing attention to a key word is to change its color from that of other words in the communication. This intensification of an aspect of a message could be developed into a hierarchy of, say, three colors whose degrees of chromatic strength indicate degrees of significance in the printed information.

Letter forms in highly descriptive words can be sliced, chopped, cut, or otherwise modified by the designer to extend the pictorial aspect of their role in graphics.

8

Alternatively, when a high pulse rate of visual contrast is justified, different colors or even different character types can be introduced into the formation of individual words.

Letter forms can also be mirrored, echoed, and reflected to achieve rhythmic or repetitious patterns that borrow their significance from the content of their meaning.

N.B.: A good method of experimenting with visual tolerances of letter form distortion is to project characters and words onto variously shaped surfaces and photograph the results.

Distorted and Perspective Lettering

1 Lettering can be enlarged or reduced by simply covering the selected letter or word with a regular grid. The letter or word is then redrawn into an enlarged or reduced grid with the same number of squares as the original grid.

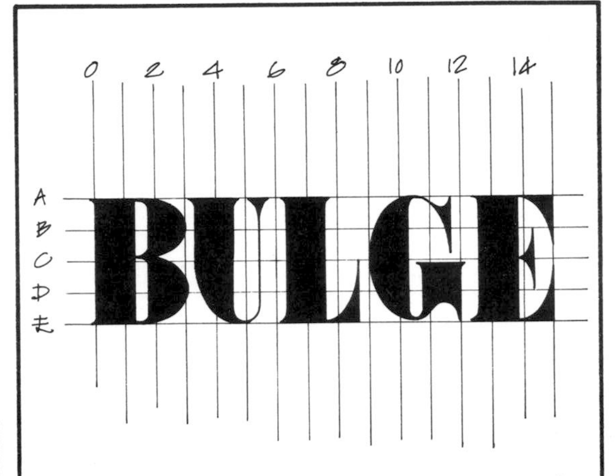

2 Also using grids, lettering can be made to distort. First, establish the letter or word, then cover it with a regular grid having its lines numbered for reference.

3 Then freehand a distorted grid, using the same number of lines as its regular version. The letter or word can now be outlined in response to the suggested distortion and, if necessary, traced for its transfer into the recipient design.

4 Words can also be transformed into perspective settings. A quick method is to project lines from a station point through a horizontal line to a baseline marked off with letter widths and spaces and angled to represent the degree of acuteness.

5 The horizontal line now becomes the baseline for the lettering in perspective. Horizontal guidelines that determine various heights should be marked on the right and extended to meet the vanishing point on the left.

6 Letters and words can be expanded, condensed, foreshortened, curved, and angled into limitless spatial positions.

Combining Lettering and Image

When words are integrated with a picture in the same format, their combined communicative powers have a double impact. One functions in a manner that the other cannot. In other words, one reinforces the other.

1

2

For combining words and pictures two basic design attitudes are outlined by the designer Emil Ruder. He suggests that one method of achieving harmony between lettering and image is to seek the closest possible formal relationship between the two.

Otherwise, the dynamics of indirect word-picture relationships can be exploited.

A second attitude is to seek a contrast between lettering and image.

N.B.: This polarity of approach is based on the pictorial confusion that often results from mixtures of these two design attitudes.

3

4

Also, direct compositional links between word and picture formation can be established.

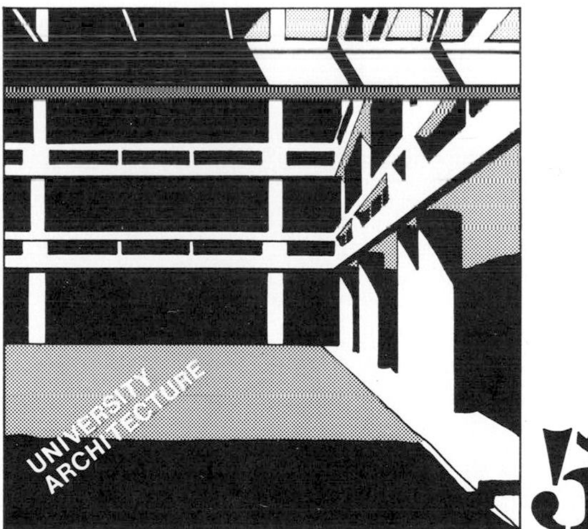

5

6

A further attitude responds to the life of the graphic image. For example, will extended usage require a sense of permanence and timelessness?

In this sense, a graphic may represent a product, a corporate image, or an institution.

Scissor-Cut Lettering and Torn Paper Shapes

Scissor-cut lettering from colored papers, though it breaks many rules of design, is very distinctive when used in graphic design. It has a genealogy rooted in modern art and was used brilliantly by such artists as Henri Matisse, Sonja Delaunay, and Stuart Davis. Here are examples of lettering by the respective artists.

1

2

The design of cut-out lettering reflects the whim of the designer. It rarely conforms to setting up lines when introduced into graphics, but can use different colors for letter forms or for words to highlight key points in the message.

3

There is no need to predraw letter forms prior to cutting. Simply fix in mind an idea of the desired shape. The ensuing cut should attempt to trace this conceptual outline.

N.B.: The cutting action should be fluid; that is, an action that carves rather than clips the shape. Matisse saw this action as conceptually releasing an imagined form from the paper. He said, "Cutting straight into color reminds me of the direct carving of the sculptor."

4

A good method of producing lettering for posters, labels, and exhibition panel headings is to cut strips of paper and glue them end down onto prepenciled guidelines drawn on a baseboard of the same or a different color. If required for reproduction, these can be photographed under a strong side light.

5

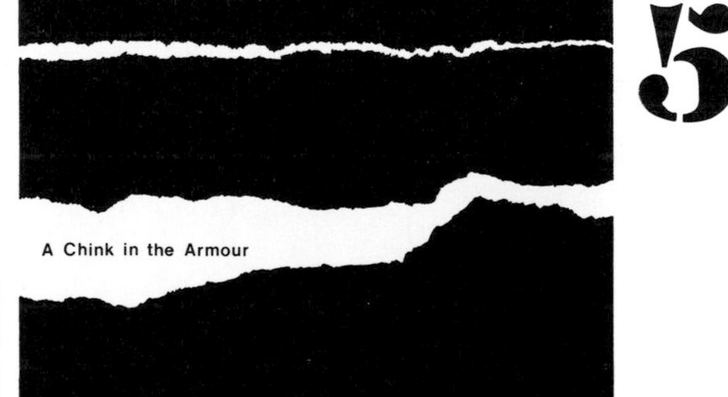

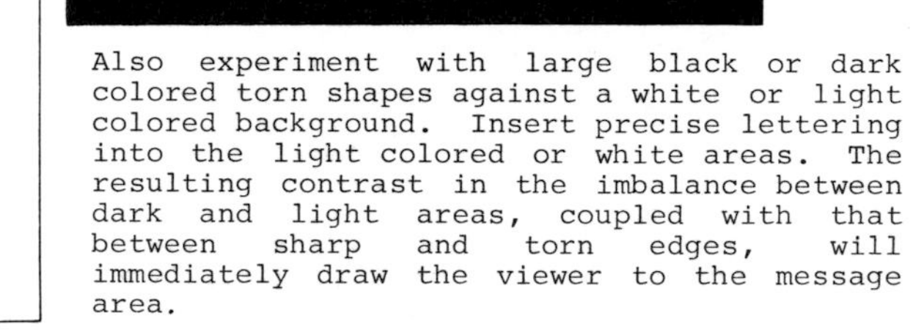

Also experiment with large black or dark colored torn shapes against a white or light colored background. Insert precise lettering into the light colored or white areas. The resulting contrast in the imbalance between dark and light areas, coupled with that between sharp and torn edges, will immediately draw the viewer to the message area.

Contrast as Graphic Ingredient

1 The human eye, like that of many lower animals, is attracted by movement and change in the visual field. This response to visual stimulation is a by-product of our evolutionary survival kit that can, through deliberate contrasting of elements, be catered to in the design of graphics.

For example, the combining of two contrasting elements, such as the roughness of a heavily textured shape with the machined precision of instant lettering or typography, can both alter and enhance the visual effect of each element. Experiments in the juxtaposing of dissonant lines, shapes, textures, and letter forms therefore provides an important reservoir of experience for the designer.

There are many forms of contrast in graphics. At a basic level, however, the tensions created between positive and negative elements is a tension that is a product of contrast, because combining like elements with like results in blandness and monotony.

Degrees of contrast or tension can be articulated by the proximity and juxtaposition of dissonant lines or lines describing the edges of two different forms.

Differences between solid and textured areas, or a variation in the positive and negative pulse rates of textures, can also extend the visual effect of contrast.

However, in exploiting contrast, take care to avoid the breakdown of the overall uniformity of a design. Extremes of contrast can allow individual forms to dominate and even complete displays of elements to be negated.

Animating the Frame of Reference

1

The journey of an image from concept to graphic statement may not finally be achieved even when all the marks have been made. Following from the artist's adage that what is excluded from an image is just as important as what is retained, the fixing of its frame of reference can be a crucial design stage. This occurs because the placement of the format's limits, or the "window" through which an image is viewed, can radically alter and, indeed, improve the impression it creates.

This procedure simply requires a cut-out paper window, or a range of paper window sizes and shapes, to determine the best picture area. It involves evaluation of the image in relation to its edge or border.

2

This stage can be filled with surprises as the apparent balance and composition of the same image will shift and change in response to different positions and shapes of the window frame.

Another method of picture area selection that is used widely in publishing prior to the cropping of photographs and artwork can be adopted. This uses two L-shaped pieces of thick board that, when placed around the graphic, allow a flexible means of determining the best picture area.

3

4

One option worth consideration is to allow part of the image to penetrate a drawn frame. Apart from the resulting visual game with space, the window also acts to frame, and therefore pinpoint, the center of attention in the graphic.

2 DRAWING AND DESIGN FOR REPRODUCTION

Line and Tone Originals

Because the graphics of a normal design process will undergo a printed transformation by diazo, photocopier, and camera, many designers have turned to the wider opportunities available in the reproduction stage. For example, some produce printed communications such as brochures, posters, exhibition panels, and promotional literature to sustain an image, while others seek wider exposure in the print media, such as journals and magazines. It is therefore the intention of this section to introduce some basic principles and techniques related to the printed format, and also to help the designer predict the effects that a drawn image can have on its printed version.

1

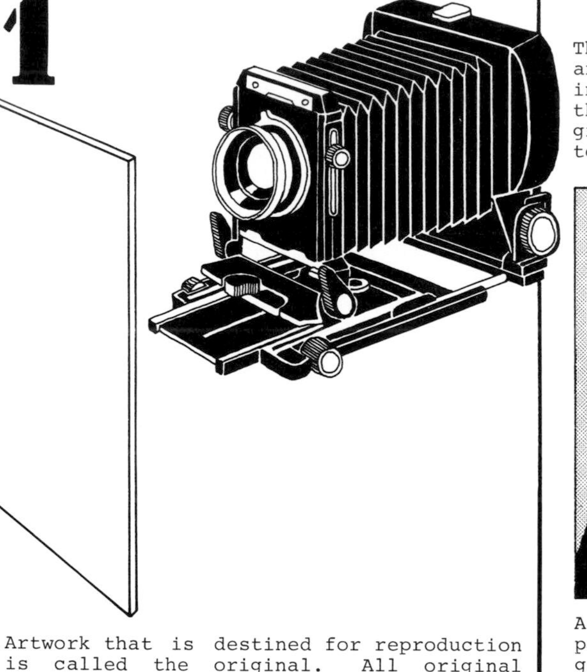

Artwork that is destined for reproduction is called the original. All original material has to be transformed via the medium of the reproduction camera into a form compatible with one of the printing processes.

2

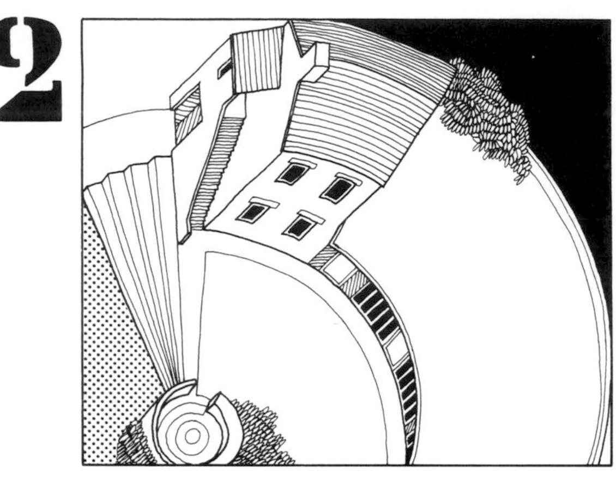

There are two basic types of original: line and tone. Line originals, such as ink drawings, should consist entirely of line work that is both solid and dense, because any gray tones in a line-shot original will tend to be lost.

4

A side effect of screening is the moiré pattern sometimes found in reprocessed photographs. This is caused when an already screened image has been reshot for further publication, the resulting effect representing a clash between the two grades of screen. To avoid the moiré pattern, the same size screen must be used in both stages.

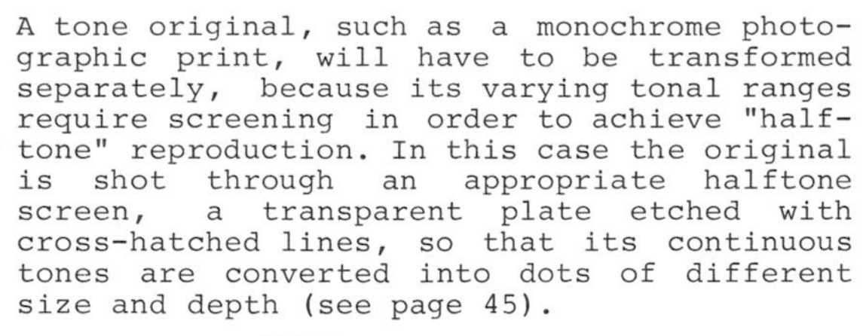

A tone original, such as a monochrome photographic print, will have to be transformed separately, because its varying tonal ranges require screening in order to achieve "halftone" reproduction. In this case the original is shot through an appropriate halftone screen, a transparent plate etched with cross-hatched lines, so that its continuous tones are converted into dots of different size and depth (see page 45).

The use of coarser halftone screens, as in newspaper publication, reveals evidence of the printed dots upon close scrutiny.

5

Color originals, whether in line or tone, require separation of their hues by means of color filters. This process translates an image, such as a painting or transparency, into its color components for reproduction via four-color process inks (magenta, yellow, cyan, black). Halftone screens may also be used at this stage or be introduced at a later point.

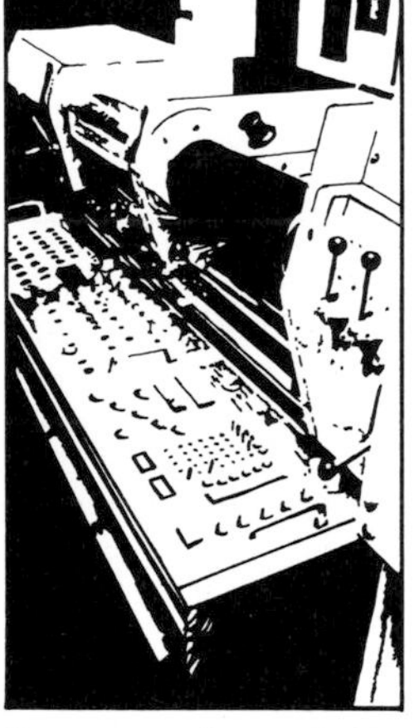

The use of laser light for scanning full-color originals is a color-separation technology that is fast replacing the process camera. The electronic scanner operates a scanning and recording drum that accepts both transparencies and nontransparent originals for scanning by laser and a color computer.

Commercial Methods of Reproduction

Letterpress is the traditional process of printing using a raised surface. Although associated with movable type, modern letterpress uses photoengraved plates for line and tone, monochrome and color reproduction. Metal or plastic plates are produced by screening a photographic negative so as to transpose the image into a constituent system of halftone dots of varying size. This achieves a full range of tonal values in the final print.

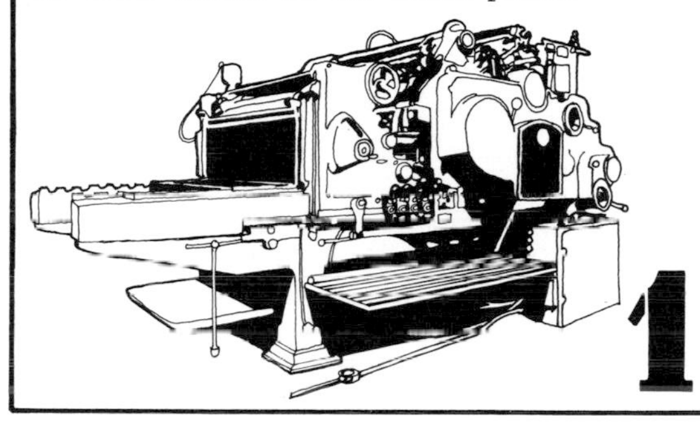

Photolithography is a planographic process, as it relies upon the mutual repulsion of water and grease on a flat plane. During printing the image attracts the ink, while the film of water protects the nonprint areas (see pages 89-96).

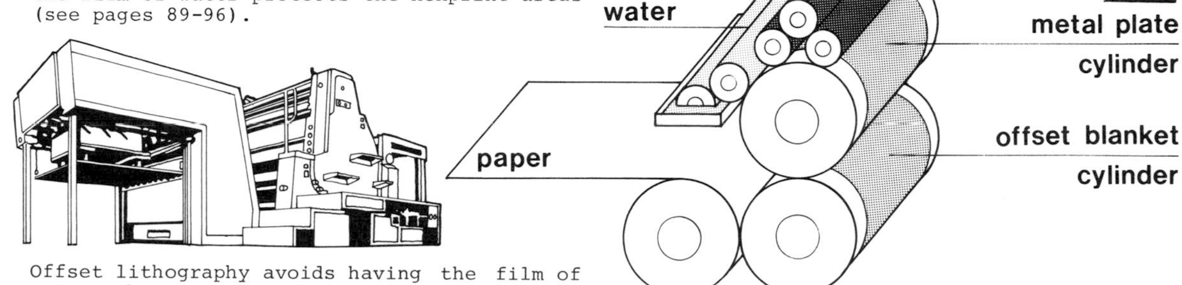

Offset lithography avoids having the film of water make contact with the paper. To do this the inked image is transferred from the photographic plate onto a rubber-coated blanket cylinder. This transfer is then carried to the paper minus the moisture.

As with letterpress, the regular lithography process prints images backward. Another advantage of the offset method is that the printed image is returned to the paper the right way around.

Photogravure is for high quality black-and-white and color reproduction. It is an intaglio process, with ink being transferred to the paper from small indented cells of different depth. The deeper the cell, the more ink it carries and the darker the impression. The printing surface can be produced by engraving, in which a cutting head indents the metal plate in response to signals conveyed from either an electronic or laser scanner that moves over the original.

N.B.: Modern scanners can achieve ready screened and sized color separations.

3

Two other processes of note are collotype and screen-process or silk-screen printing. Collotype is the only method of printing that produces a truly continuous tone, making it ideal for the reproduction of fine artwork. Like lithography it is a planographic process, in which a film of gelatin carries the image. However, it is a costly and slow medium and is used only for small runs of high quality work. Screen process printing is a form of mechanized stencilling, with ink being forced through a fine mesh of metal or fabric stretched over a frame. Hand-cut stencils have a limited life, whereas photographic stencils can give up to 10,000 impressions (see pages 72-87).

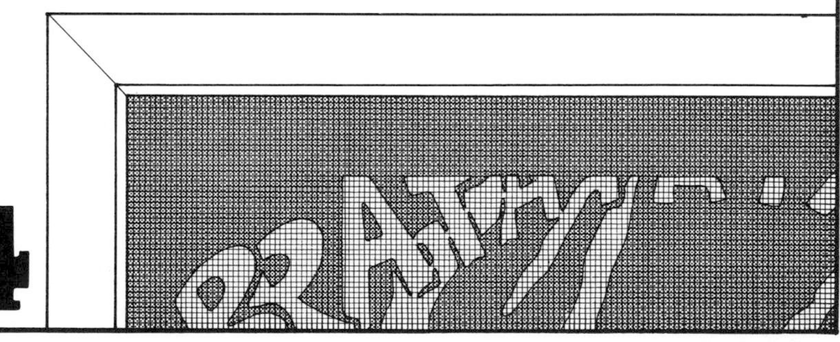

4

Effects of Enlargement, Same-Size, and Reduction

1 Generally speaking, enlarging a drawing as a print does not improve its inherent quality. Reproduction on a larger scale will amplify imperfections, fragment lines, erode crisp edges, and disrupt tones and patterns. The later repair of these blemishes by retouching can also be an expensive operation.

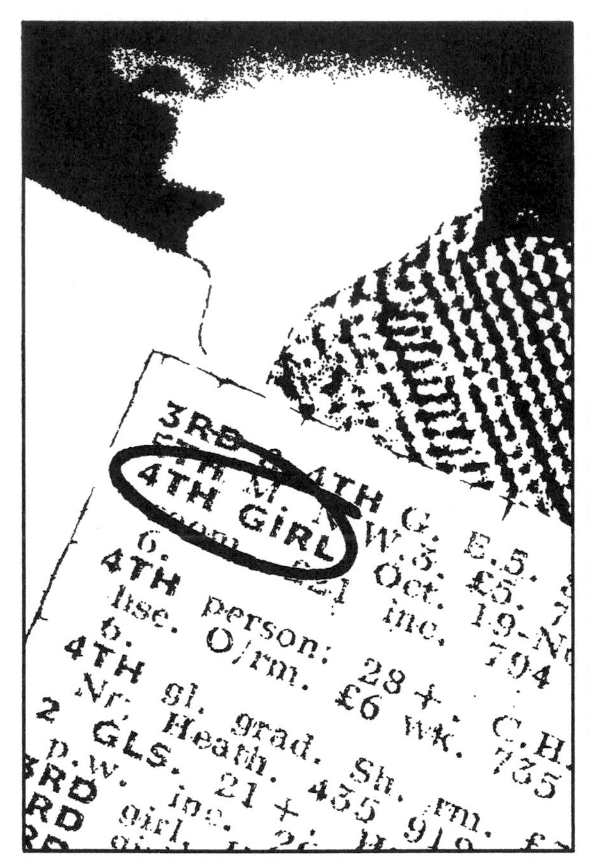

However, there are occasions when the apparent charm of these imperfections is sought by the designer. For example, in exhibition and display panels, drawings, engravings, lettering, and so on are enlarged specifically to achieve this effect.

2 The same can be said for the reproduction of images at same size (one to one). In this instance the printed version is only as good or as bad as the original material.

4 Whenever possible, all artwork destined for reproduction should be produced to a consistent size. Apart from affording the designer greater control over the consistency of the printed artwork, having original artwork of a consistent size also reduces costs, because the reduction control on the reproduction camera has to be set once only for shooting a complete set of originals.

3 This is why the majority of designers usually employ the method of producing artwork "half up" or "twice up" (see page 41) from the size intended for reproduction.

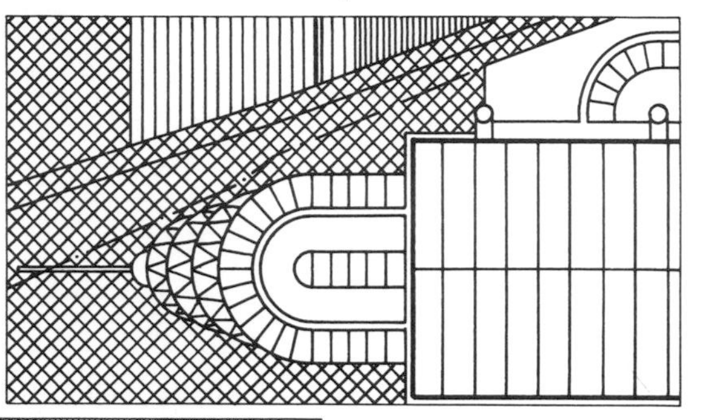

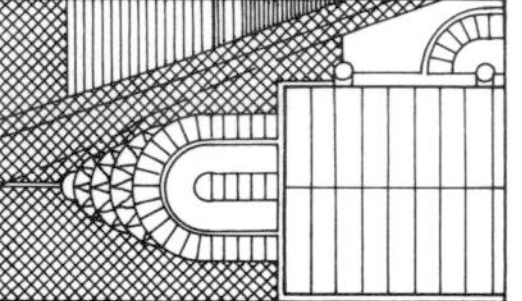

In this case the printed version improves by becoming a denser, sharper image in which any blemishes tend to disappear.

5 It is important therefore to preplan a set of drawings for reproduction. Decisions should initially be made concerning the amount and degree of detailed information to be communicated.

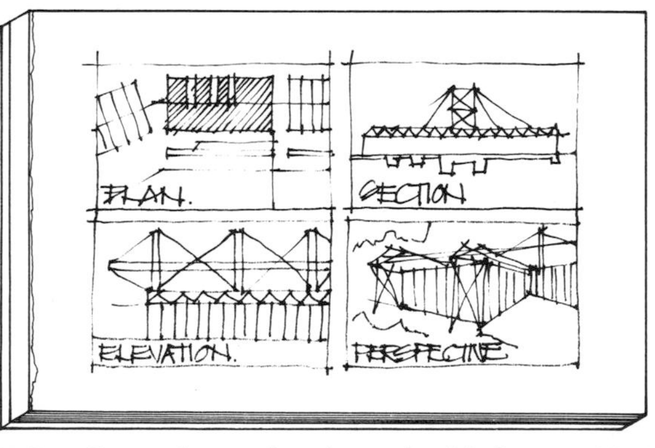

N.B.: Excessive reductions should be avoided because the original will be large, awkward to control, and difficult to process.

Image Reduction and Proportional Formats

When artwork is reduced, its format area becomes smaller while its width and height dimensions, though diminished, remain proportionally the same. However, when designers speak of producing originals "half up" and "twice up," they refer to linear rather than area scaling. For example, a twice up or 2 : 1 ratio gives a reduced format area of one-quarter that of the original. Similarly, a 3 : 1 reduction is one-ninth of the original area, and so on.

2 This is the basis of the "A" series format, an international standard of sheet sizes whose sides are in the ratio of 1 : √2. No matter how many times the sheet is folded in half, each subdivision retains the same ratio. This facilitates reduction and enlargement of printed matter, diagrams, and drawings by photographic means. These sizes are in use wherever the metric system is adopted. In the near future they should become available in the United States.

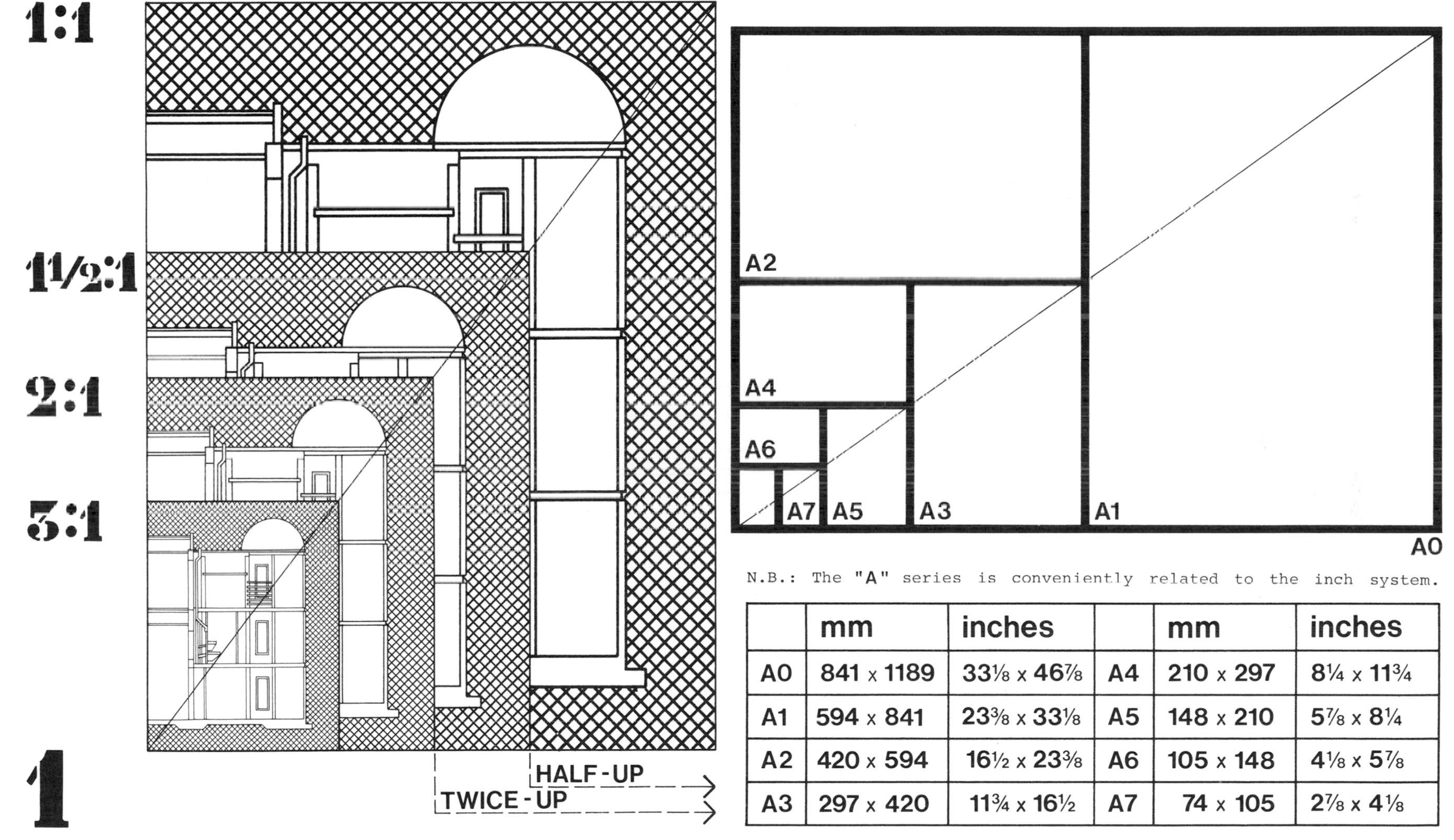

N.B.: The "A" series is conveniently related to the inch system.

	mm	inches		mm	inches
A0	841 x 1189	33⅛ x 46⅞	A4	210 x 297	8¼ x 11¾
A1	594 x 841	23⅜ x 33⅛	A5	148 x 210	5⅞ x 8¼
A2	420 x 594	16½ x 23⅜	A6	105 x 148	4⅛ x 5⅞
A3	297 x 420	11¾ x 16½	A7	74 x 105	2⅞ x 4⅛

Image Reduction: Line

1

Illustration drawing covers a wide variety of image types, from architectural orthographics and technical diagrams to perspective drawings and freehand artwork. However, the decision to prepare artwork for reproduction at a size larger than that intended in the reduction will depend to a great extent on the degree of complexity within the image.

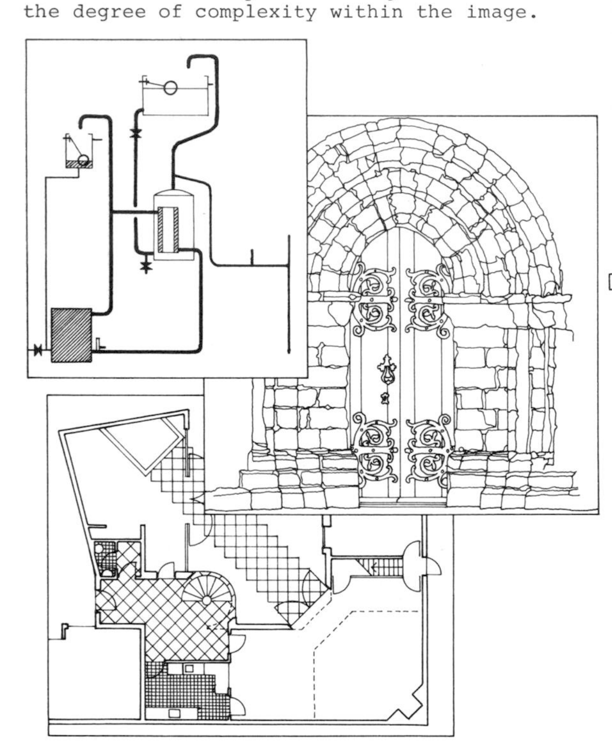

When producing artwork for a line shot by the reproduction camera, it is important to draw on a smooth, hard surface, such as good quality art board, cartridge paper, or tracing paper. It is this care, together with the quality of the print paper, that plays such an important role in helping images survive transformation by reduction.

2

Although the camera responds to dense and contrasting images, make sure that preparatory drawings, such as construction lines and the like, are nonreproducible. These should be worked either in a faint, hard-grade graphite pencil or in the light blue, nonprinting or "nonrepro blue" pencil.

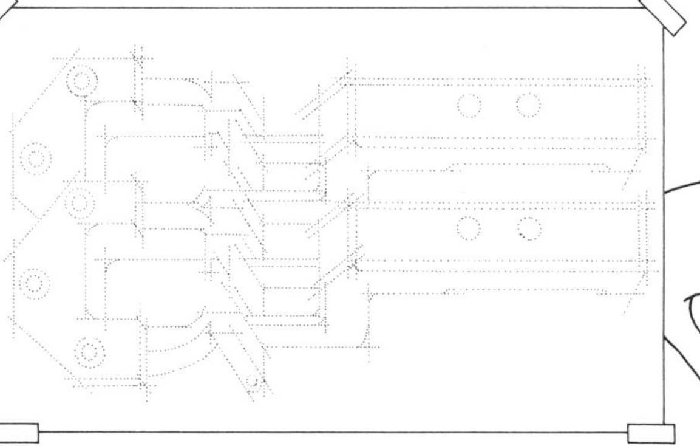

Being more susceptible to erosion by reduction, the thinner lines are the most critical. For this reason their weight should be determined at the outset, and on the basis of their intended reduction ratio.

A fine line weight can be established by calculating backward from its intended thickness when printed. For instance, if a fine line is to exist in printed form at 0.1mm, then the line, prior to a 2 : 1 reduction, should be drawn to a thickness of 0.2mm. The same reduced line thickness should be drawn at 0.3mm prior to a 3 : 1 reduction. From this fine line, the other line weights can be determined. The accompanying table is offered as a basic framework. However, if in doubt, always tend toward a line thickness of a bolder size.

REDUCTION	FINE LINE	INTERMEDIATE LINE	HEAVY LINE
3 : 2	0.15mm	0.3mm	0.6mm
2 : 1	0.2mm	0.4mm	0.8mm
3 : 1	0.3mm	0.6mm	1.2mm

As contrast and clarity are salient ingredients in reproduction, it is vital to structure images from a discernible hierarchy of line weights. A basic hierarchy may comprise three or four line thicknesses, from fine to heavy.

N.B.: Depending on the type of drawing, line weights may be organized in relation to a hierarchy of information, or by the spatial position they occupy in the field of view.

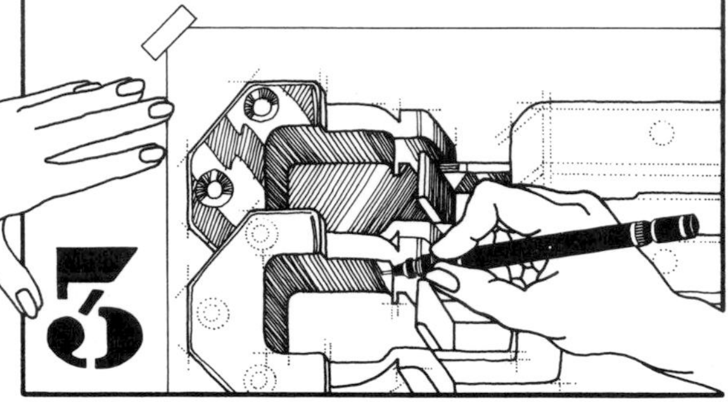

3

4 5

A further factor in reduction is the allowance made for intervals between lines. A common error, for example, is to draw a series of lines or hatching in close proximity which, when reduced, will tend to fill in.

The basic rule is, therefore: The larger the reduction, the wider the spacing--and never draw lines with less than their own thickness as the interval between them.

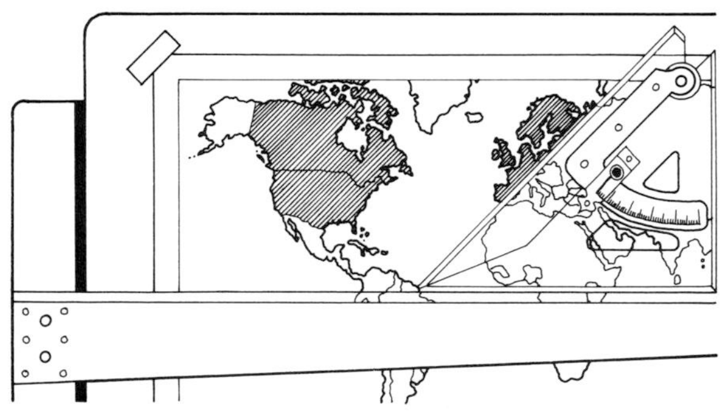

Image Reduction: Mechanical Tone

1

Sheets of mechanical dry-transfer screens and symbols can be used to good effect when introduced as tones and as elaboration into line drawings and diagrams. The screens are available in different shades of dots and lines. The range of density in sheets of dots is specified as a percentage of solid black and as the number of lines of dots to the inch, with the bigger the dots, the fewer lines of dots to the inch. Similarly, line screens are specified by their number to the inch and come in various line weights.

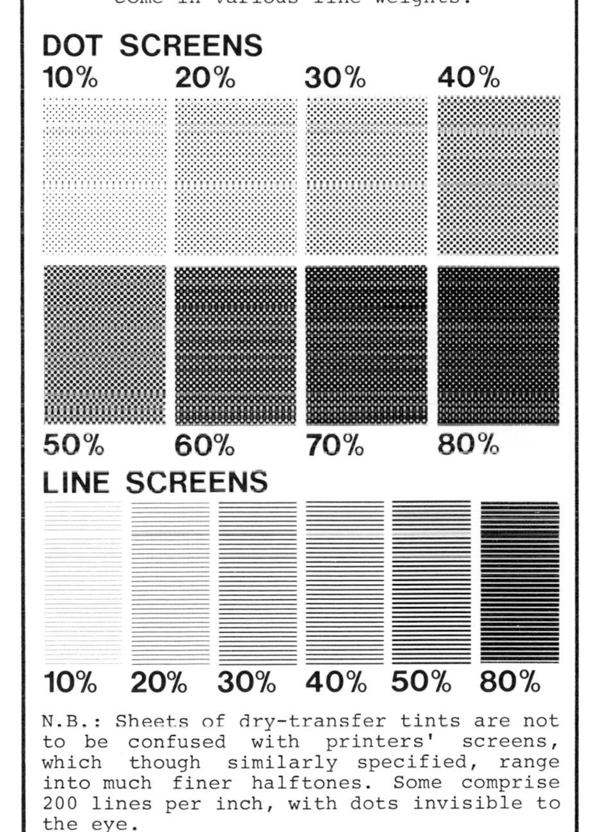

N.B.: Sheets of dry-transfer tints are not to be confused with printers' screens, which though similarly specified, range into much finer halftones. Some comprise 200 lines per inch, with dots invisible to the eye.

2

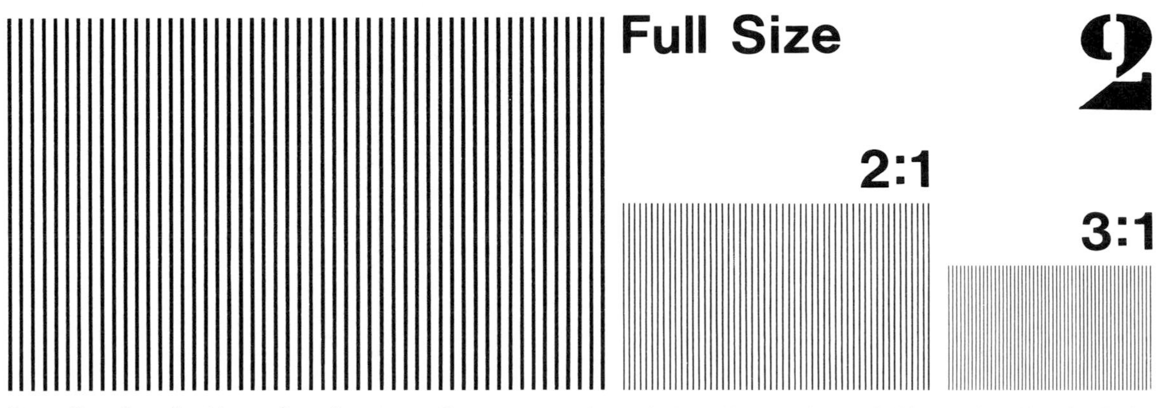

As a 2 : 1 reduction of a dry-transfer screen gives twice the number of lines to the inch in its printed version as a same-size print, and a 3 : 1 reduction gives three times the number of lines, it is a good idea to see a catalog of printers' screen tints before using the press-on variety. This experience would enable the designer to predict the effects of reduction before applying screens to the artwork.

N.B.: Because the number of lines on a screen is multiplied by the amount of the reduction ratio, it is worth avoiding the darker end of the range, for instance, 70 and 80 percent, as they tend to fill in when reduced and appear for all intents and purposes as black.

3

It is best to structure steps of positive contrast between various grades and densities of screens on the original artwork, since a close similarity between screens at full size will appear to look alike in its reduced form.

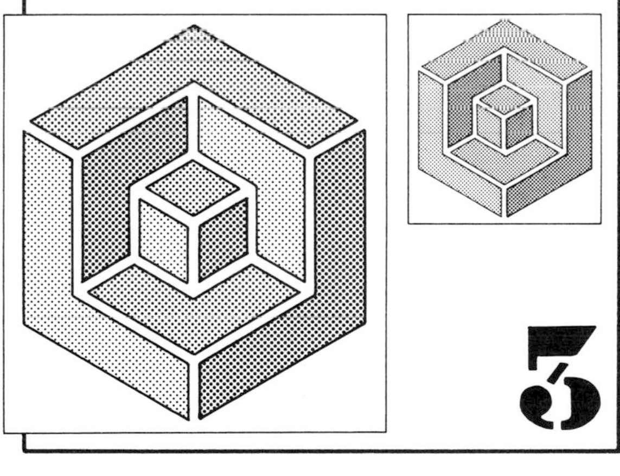

4

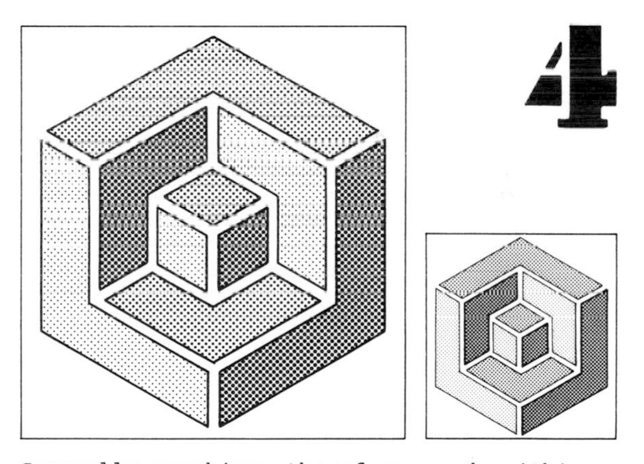

Generally speaking, therefore, work within a basic range of clearly stepped densities of, say, 10 to 60 percent and between 27.5 and 60 lines per inch. Always apply dry-transfer screens to smooth, stable, drawing surfaces, preferably artboard, and, to avoid a breakdown in legibility, never apply screens across small type or fine lines.

Hints when Preparing Artwork in Line

1

The main consideration for the beginner when producing line drawings for reproduction is to use a sharp, dense black line on white board.

N.B.: If you are drawing on transparent material, avoid working on the back of the sheet to achieve value differences in the artwork. Also, do not use colored lines. Both these techniques can create problems during reproduction.

2

In order to guarantee a predictable standard of printed reproduction it is recommended that more intricate artwork, such as architectural and technical illustrations containing fine lines and dense hatching, be prepared for same-size, or one to one, printing. The lens of the printer's process camera performs best when it is shooting for one-to-one sizing.

To avoid any chance of incompatibility between the size of the original and the printer's copyboard (the easel that carries the original for sizing and shooting), it is worth checking this dimension before commencing the drawing.

When preparing artwork for reduction or enlargement, it is also worth checking the resizing capability of the printer's process camera. This is usually between 20 and 250 percent. Once the range is established, however, you should always work within it.

N.B.: The precise resizing of originals is obtained by adjusting the camera lens to film distance and also the camera to copyboard distance, both settings being arrived at by sliding these components along a track.

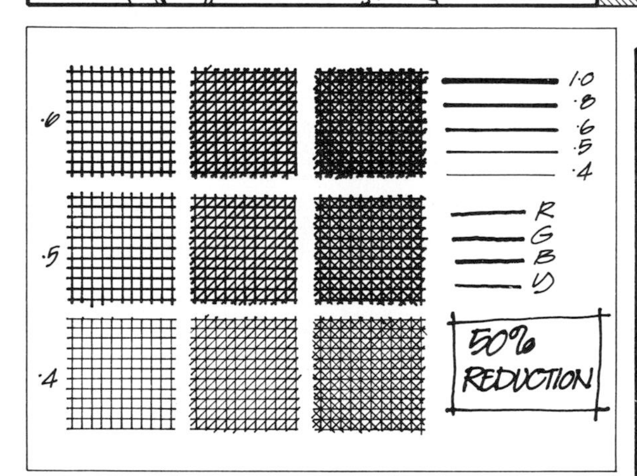

4

If a reduction is required, it is worth producing a test card showing an annotated range of line weights plus a variety of hatching densities and, possibly, some colored lines. Then ask your printer to make a trial photographic reduction.

5

The resultant contact print will, at the given reduction, provide an accurate reading of the potential of each line prior to the production of the drawing. It will also act as a guide to help select appropriate lines for future artwork.

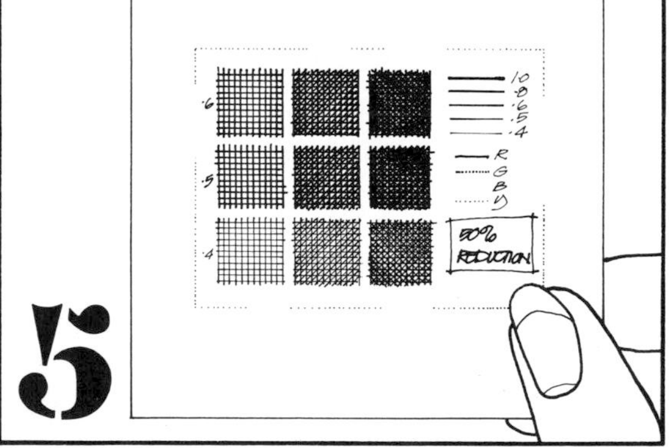

Hints when Preparing Artwork in Tone

1

When preparing tonal artwork for reproduction, such as drawings in graphite, conte crayon, charcoal, watercolor, and pen and wash, it is important to eliminate the background paper within the area occupied by the drawing. In other words, to ensure that the integrity of the original is maintained as an isolated tonal image on the printed format in drawings for reproduction, highlights that would appear as unrendered white paper in a regular drawing should be rendered in a degree of tone.

A An enlargement of a halftone screen.
B An enlargement of a halftone print.
C A cross-section of a halftone block showing graduation of dots from near solids to highlights.

The reason for this additional tone is that problems can occur when the printer shoots the original through the halftone screen. During this transformation of the image, areas of a lighter tone in the original will be subsequently converted into tiny dots in the printed version. Also, on those areas where the white of the paper that carries the artwork occurs, no dots should appear. Therefore, in order to help the printer achieve this separation, there should be a sufficient value contrast between the white of the paper and the tone of the lightest area contained within the artwork.

2

This degree of contrast, called also step-off, is obtained when highlights (or the lightest parts of the drawing) are rendered approximately 10 percent darker than the white of the paper.

N.B.: In representing the lightest sections of the drawing, this degree of highlight value will naturally readjust the scale of the progressively darkening values contained in the drawing.

3

4

A point worth remembering is that you can always request your printer to lighten the value of highlights during the process of reproduction. This is a more satisfactory method than the alternatives forced upon the printer when having to compensate for artwork with inadequate tonal separation. However, if in any doubt, it is a good idea to ask the printer for a trial halftone transformation of the original. This preview will illustrate the degree of highlight tone required by the reproduction process and thus avoid any degradation of the printed image.

N.B.: When supplying tonal artwork, it is always better to submit the original rather than a photographic positive. In most cases, prints cannot provide the tonal separation required for a high quality reproduction.

Typeface: Points and Picas

1 Visual consistency and appropriateness are important considerations when selecting a typeface. A good discipline for beginners is the use of a single typeface throughout a basic design. Such a controlled difference in the size and weight of typeface will communicate clear visual signals. It is also important to select an established face rather than one that may be fashionable, as the latter could quickly date. These faces are some of the classic typefaces that are also available in dry-transfer form.

Modern No.20
ABCDEFGH
IJKLMNOP
QRSTUVWZ
XYabcdefg
klmnopqrs
wxyz12345
890ÆŒ?!%

Baskerville Old Face
ABCDEFGHI
JKLMNOPQ
RSTUVWX
abcdefghijk
nopqrstuvw
1234567890
Æ&£$?!(;;)

Gill Sans
ABCDEFGHIJK
LMNOPQRST
UVWXYZabc
ghijklmn
vwxyz
567890?
EØ&ß$

Helvetica Medium
ABCDEFGHI
JKLMNOPQ
RSTUVWXY
Zabcdefgl
lmnopqrs
wxyz1234
7890ß&?!

Times Bold
ABCDEFGHI
JKLMNOPQ
RSTUVWXY
bcdefghijkl
opqrstuvwx
234567890
E$£&?!%;«»

Univers 45
ABCDEFGHIJK
LMNOPQRSTU
VWXYZabcde

2 The measurement of a typeface is based on the point system, the factor of which is one point or .0138", which is approximately 1/72" (0.35mm). Seventy-two points equate closely enough to one inch to be accepted as such. Standard type sizes therefore are multiples of these seventy-two points.

0 3 6 9 12 16 18 PICAS

8 POINT TYPE

J'étais arrivé au sommet d'un coteau. Avant de descendre sur le revers je jetai encore une fois les yeux sur la cure, que j'allais perdre de vue Le soleil, près de se coucher, dorait d'une lisière de pourpre la crête des tilleuls et le sommet des vieilles ogives du presbytère, tandis qu'une

12 POINT TYPE

J'étais arrivé au sommet d'un coteau. Avant de descendre sur le revers, je jetai encore une fois les yeux sur la cure, que j'allais perdre de vue. Le soleil, près de se coucher, dorait d'une lisière de

16 POINT TYPE

J'étais arrivé au sommet d'un coteau. Avant de descendre sur le revers, je jetai encore une fois les yeux sur la cure, que j'allais perdre de vue. Le so

24 POINT TYPE

J'étais arrivé au somme t d'un coteau. Avant de descendre sur le revers je jetai encore une fois

The pica is an Anglo-American measurement that approximates 1/16" (4.2mm). It is used to specify the width and depth of areas of type, such as columns, irrespective of the size of type in use. When illustrations are incorporated into blocks of text, they can also be specified in picas.

N.B.: Six picas equal 72 points, or 1" (25.3mm).

Typeface: More Points

1

Having established that 12 points equal 1 pica, it is worth adding that another, unrelated system of typeface measurement exists. This is the didot point system. It is slightly larger than the Anglo-American point and refers to typefaces originating from many European countries. The designer therefore has to be cautious when specifying type, because a French-designed work in the same point size as, for example, one designed in the United States will be larger. This illustrates how the two systems differ:

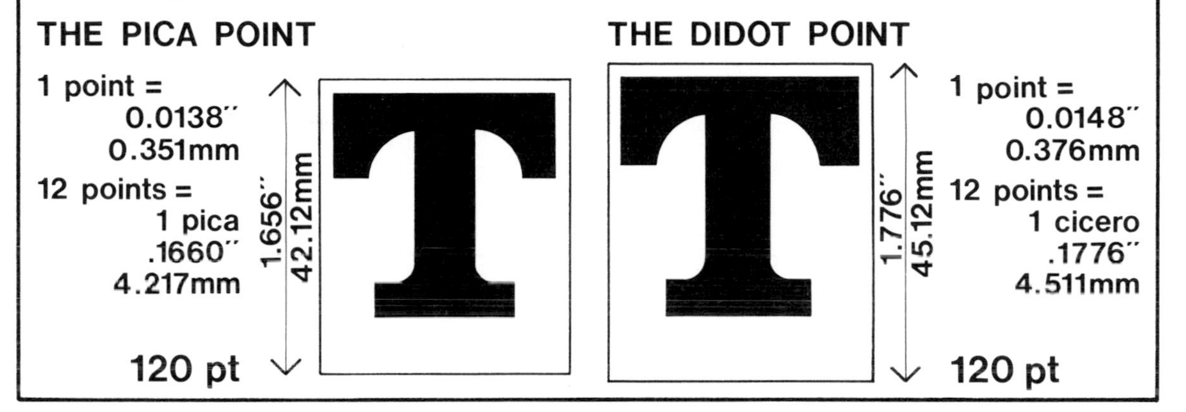

2

This is the basic anatomy of letters set in type:

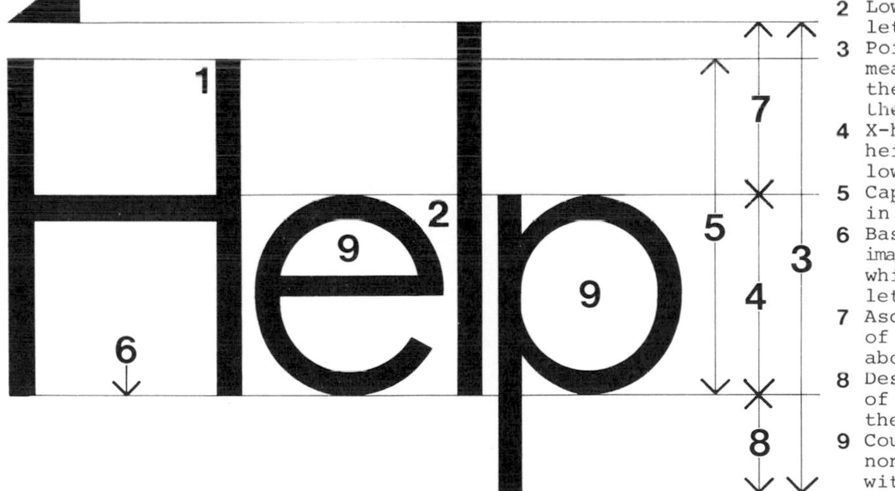

1 Capital letter, or uppercase (sometimes called a majuscule).
2 Lowercase, or minuscule letters.
3 Point size: The measurement taken from the top of the **l** to the bottom of the **p**.
4 X-height: The overall height of the body of lowercase letters.
5 Capital height: For use in capital alphabets.
6 Baseline: An imaginary line on which the bodies of letters rest.
7 Ascenders: Those parts of a character rising above the body.
8 Descenders: Those parts of a character below the baseline.
9 Counters: The white, nonprinting areas within a piece of type.

3

Like so many of the typographical terms, leading (pronounced "ledding") stems from handset letterpress and refers to the amount of space set between lines of type. When more space is required between lines than that given by the beard (the space normally occupied by a descender) and the nonprinting area above the X-height, an extra strip of metal or lead is inserted between the lines (see page 49).

N.B.: Capitals may appear crowded if not well leaded, because they occupy a greater part of the depth of the type.

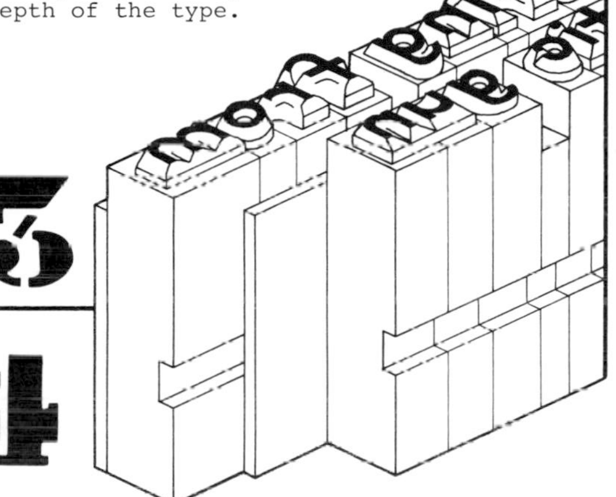

4

The traditional handset letterpress method of printing from raised type, together with its mechanical successor, hot metal (that is, the process of casting metal type in molds as each letter is typed on a keyboard), is fast being replaced by a modern strike-on technology. For instance, computerized phototypesetting uses a computer's memory to produce paper negatives for paste-ups that are developed in the same way as photographs. Also, word processors that link a computerized typewriter with a television screen produce magnetic tapes or disks for playback through phototypesetting machines, creating instant copy. Another strike-on method is the digital typesetting machine. This is a developing technology that can instantaneously translate copy and illustrations into electronic codes capable of sending completed layouts for printing.

Hints when Designing with Typefaces

Typefaces come in different categories. A frequent distinction is made between a typeface suitable for printing large blocks of text, called book faces, and those known as display faces, which are more suitable for headings, titles, and so on.

DISPLAY

Another distinction concerns book faces, classified as either the Roman book faces or sans serif typefaces. The former draw their inspiration from the writings of antiquity, while the latter have none of the serifs of the Roman type.

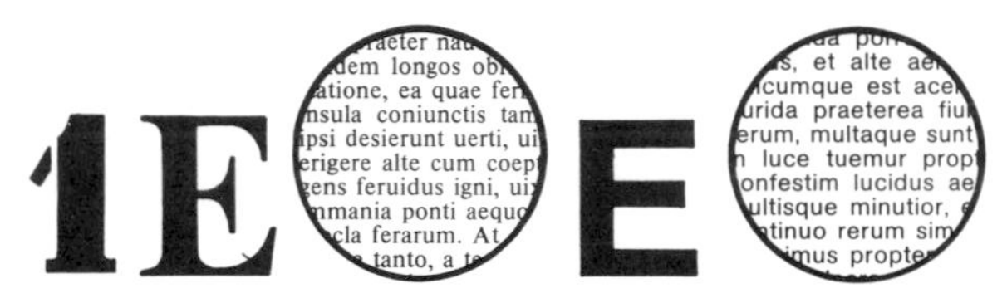

Each typeface is available in a variety of sizes that can range generally from 5 to 100 points. Most typefaces come in roman and italics. (The former is the standard face, not to be confused with the particular alphabet.) Many of the classic typefaces also offer a range of weights, such as light, medium (standard), bold, and extra-bold, together with expanded and condensed versions.

2

The graphic designer Massimo Vignelli suggests that, apart from the selection of a single typeface, a further aid to a consistent typographical appearance in printed communication is to restrict typefaces to three distinctly different sizes. Aside from making the job of typesetting easier and cheaper, it also helps the function of reading.

4

Also, the number of type weights should be limited to two. As with the typeface size, the eye finds it difficult to discriminate between more than this number during the act of reading.

N.B.: Aim for a clear distinction between the two selected weights and the three selected sizes.

5

There are three basic methods of organizing columns of text into blocks, such as those formed by paragraphs: the skipped line, the indented line, and the skipped line plus indent.

descendre sur le revers, je jetai encore une fois les yeux sur la cure, que j’allais perdre de vue. Le soleil, près de se coucher, dorait d’une lisière de pourpre la crête des tilleuls et le sommet des vieilles ogives du presbytère, tandis qu’une ombre bleuâtre couvrait de ses teintes tran quilles le vallon qui me séparait de ces lieux. A la fraîcheur du soir, l’herbe redressait sa tige, les les yeux sur la cure, que j’allais perdre de vue.

J’étais arrivé au sommet d’un coteau. Avant de descendre sur le revers, je jetai encore une fois les yeux sur la cure, que j’allais perdre de vue. Le soleil, près de se coucher, dorait d’une lisière de pourpre la crête des tilleuls et le sommet des vieilles ogives du presbytère, tandis qu’une ombre bleuâtre couvrait de ses teintes tran quilles le vallon qui me séparait de ces lieux. A la fraîcheur du soir, l’herbe redressait sa tige, les

descendre sur le revers, je jetai encore une fois les yeux sur la cure, que j’allais perdre de vue. Le soleil, près de se coucher, dorait d’une lisière de pourpre la crête des tilleuls et le sommet des vieilles ogives du presbytère, tandis qu’une ombre bleuâtre couvrait de ses teintes tran quilles le vallon qui me séparait de ces lieux. A la fraîcheur du soir, l’herbe redressait sa tige, les les yeux sur la cure, que j’allais perdre de vue.
  J’étais arrivé au sommet d’un coteau. Avant descendre sur le revers, je jetai encore une fois les yeux sur la cure, que j’allais perdre de vue. Le soleil, près de se coucher, dorait d’une lisière de pourpre la crête des tilleuls et le sommet des vieilles ogives du presbytère, tandis qu’une ombre bleuâtre couvrait de ses teintes tran quilles le vallon qui me séparait de ces lieux. A la fraîcheur du soir, l’herbe redressait sa tige, les vieilles ogives du presbytère, tandis qu’une

descendre sur le revers, je jetai encore une fois les yeux sur la cure, que j’allais perdre de vue. Le soleil, près de se coucher, dorait d’une lisière de pourpre la crête des tilleuls et le sommet des vieilles ogives du presbytère, tandis qu’une ombre bleuâtre couvrait de ses teintes tran quilles le vallon qui me séparait de ces lieux. A la fraîcheur du soir, l’herbe redressait sa tige, les les yeux sur la cure, que j’allais perdre de vue.

  J’étais arrivé au sommet d’un coteau. Avant descendre sur le revers, je jetai encore une fois les yeux sur la cure, que j’allais perdre de vue. Le soleil, près de se coucher, dorait d’une lisière de pourpre la crête des tilleuls et le sommet des vieilles ogives du presbytère, tandis qu’une ombre bleuâtre couvrait de ses teintes tran quilles le vallon qui me séparait de ces lieux. A la fraîcheur du soir, l’herbe redressait sa tige, les

However, Vignelli suggests that the skipped line combined with a typeface linespaced as tightly as possible (that is, "set solid") achieves a sharpness of the left-hand column edge as well as provides a single, clear reading signal between blocks of text.

Hints when Designing with Typefaces

6

One advantage of type that is set solid is that it provides more type per area of text. It is based on the closest possible linespacing appropriate to the typeface in use and is achieved when the distance between the lines on which the typeface is set is equal to the overall point size of the typeface.

N.B.: In lowercase lettering, a minimum linespacing should avoid collision between the descenders and the ascenders.

abcdefghijklmno
pqrstuvwxyz
ABCDEFGHIJ
KLMNOPQRS
TUVWXYZ
1234567890

PLANTIN

7

Some designers find that minimum linespacing appears too compact visually. They will specify the amount of leading required. Linespacing therefore is the height of the type plus the amount of leading.

This is 14-point type
without leading.

This is 14-point type
with 3-point leading.

This is 14-point type
with 6 point leading.

N.B.: When specifying type smaller than 12 points, never exceed 2-point leading, as a wider linespacing would appear overspaced and weak visually.

unadjusted
letterspacing

adjusted
letterspacing

9

Whenever a typeface is increased in size, such as for headings and titles, a similar increase in its letterspacing will appear disproportionate and excessive. Thus it is important that the space between such letters be decreased accordingly, to maintain the consistent appearance of normal spacing.

Apart from the skipped line and the indent, the subhead is another means of defining blocks of text. In this case it is recommended that, if used consistently, subheads should exist simply as a bolder version of the same size and style of typeface. In this way, one clear signal is given to the reader.

8

capped on ten occasions for his country in a variety of positions at full international and Olympic level. Aged 40.

Stars to watch:

Zbigniew Boniek is a brilliant if slightly erratic powerhouse in midfield. Fine timing

When titles or headings are set in all-capital letters they can lose impact, especially when serifed faces are used, because they tend to join visually at the top and bottom. Possibly the best solution is to use capitals and lowercase, as this is easier to distinguish.

ALL-CAPITAL
SERIF HEADING

ALL-CAPITAL
SANS SERIF HEADING

Capital & Lower-case Heading

10

Do-It-Yourself Typefaces: Typewriters

Because of the high costs involved in setting type, the use of an electric typewriter face (as used in this manual) offers the designer a relatively inexpensive alternative. Typewriters can provide a reasonable quality of face and, with the advent of the interchangeable golfball head, a limited amount of choice. Also, recent developments begin to offer different weights and even computerized editing facilities.

1

There are three basic ways of organizing text into columns, each being achieved automatically by the typesetter and, with varying degrees of difficulty, each being capable of simulation by the typewriter.

Flush left, ragged right is natural to the typewriter, but care should be taken to create an evenness of rag along the right-hand edge. Hyphenated word breaks therefore should be avoided, to maintain the seeming randomness of this edge.

Flush left, flush right, or fully justified columns, achieves a crisp alignment of column edges. The more tightly packed effect is produced at the expense of erratic word spacing and word breaks.

Centered typography, or its typewritten simulation, is not recommended except for very special effects. Apart from the time spent in its assembly, it departs from the visual alignment of organized columns.

3

The text in this manual is justified; that is, flush left, flush right. This was not achieved automatically but as the result of an arduous process of assigning copy to a given amount of letter spaces on each line. The text was manually assembled around adjusted spacing between words and the occasional hyphenated word breaks.

1 LINE = ½ SPACE
2 LINES = 1 SPACE
3 LINES = 1½ SPACE
1234567890123456789012345678901234

The//text//in//this//manual//is********
justified;//that/is,//flush/left,***
flush right. This was not achieved
automatically but as the result of
an//arduous/process/of//assigning***
copy to/a/given amount/of/letter**
spaces on each line.//The text was*
manually assembled around adjusted
spacing//between//words//and//the*****
occasional hyphenated word breaks.

4

Typewriter characters come in two basic sizes: elite and pica. In elite type there are twelve characters to the inch, in pica ten characters.

pica pica pica pica pica pica pica p

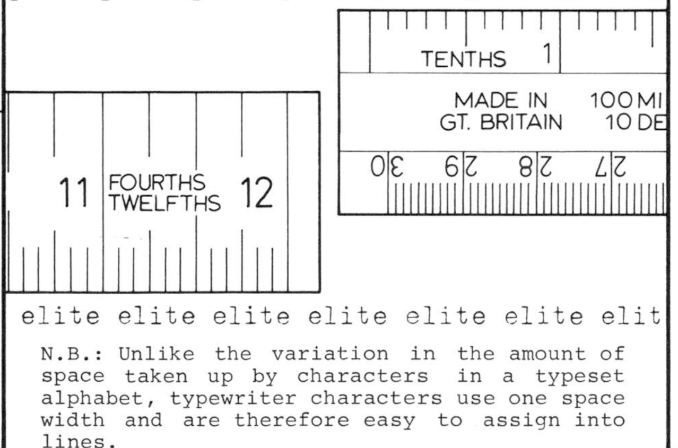

elite elite elite elite elite elite elit

N.B.: Unlike the variation in the amount of space taken up by characters in a typeset alphabet, typewriter characters use one space width and are therefore easy to assign into lines.

5

Once typewriter copy is reduced by 15 to 25 percent, it becomes comparable in size to normal text types. It also achieves a more typeset appearance, because its irregularities diminish and an overall even impression is increased.

Once typewriter copy is reduced by 15 to 25 percent, it becomes comparable in size to normal text types. It also achieves a more typeset appearance, because its irregularities diminish and an overall even impression is increased.

Do-It-Yourself Typefaces: Other Methods

Although hand lettering has been successfully applied as text in some publications, it is not recommended unless the designer has time on, and confidence in, his or her hands. Together with stenciled lettering, hand-drawn lettering is best used for notes, annotation, and labels in drawings.

A highly professional appearance can be achieved with dry-transfer lettering, however. Although possible, this would prove an arduous task if used to set text, but it is useful for headings, subheads, titles, and labels.

When using dry-transfer lettering, set the characters on a baseline drawn in non-printing blue pencil. Also, use its spacing system, especially with the larger alphabets. The smaller letters are easier and quicker to apply, and experience will allow you to space these by eye.

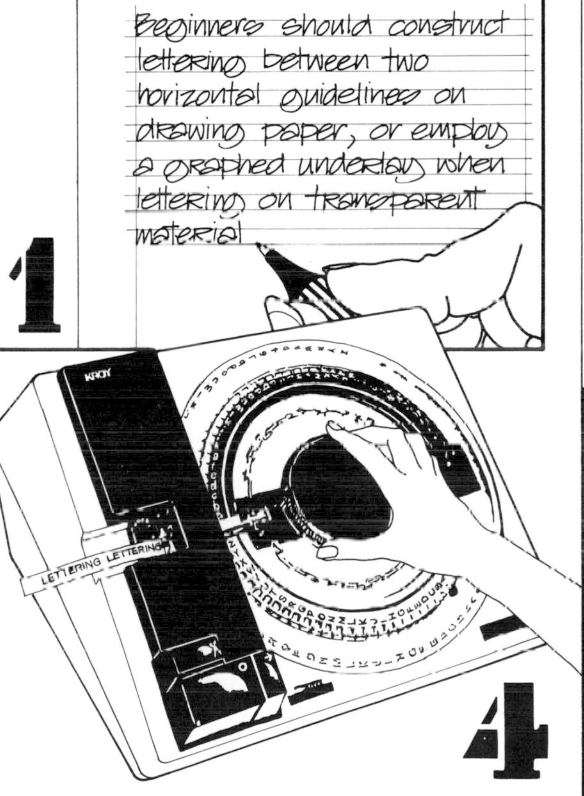

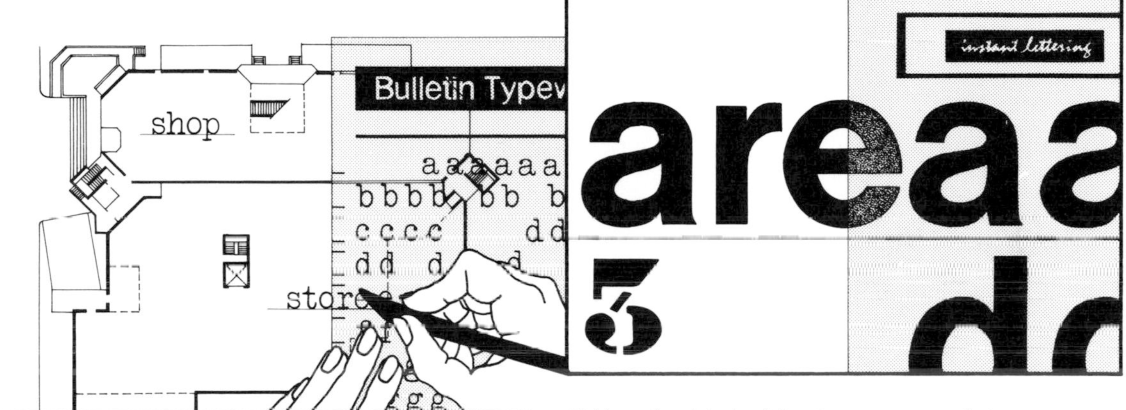

Type disks come in a host of typestyles offering black, white, and colored lettering in lowercase, uppercase, numerals, and symbols. Alphabets are generally available in 8-36 point (2-9mm) characters.

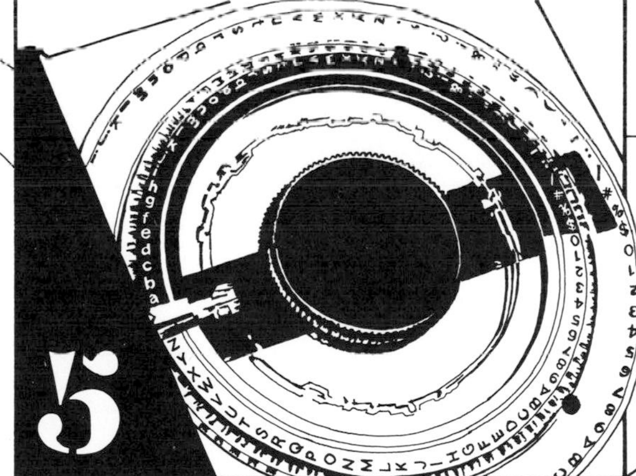

Although obtainable in opaque and transparent form, the photo-quality clear tape is best for typesetting. When printed, the special adhesive on the back of the tape is exposed by removing its backing paper. The lettering can now be attached to drawings or paste-ups, its adhesive allowing for repositioning if necessary.

N.B.: The sharp definition of the type from these machines can achieve legibility at up to fifty times in reduction.

A viable alternative is the lettering machine, which is an automatic method of setting dry-transfer lettering. This electrically powered studio printer, when loaded with a clip-on type disk, operates by dialing the character and pressing its single key. This action produces printed strips of precision-spaced instant lettering.

Grids as Layout Aids

Grids are important design aids in the creation of consistent and accurate layouts, as they carry features common to all repetitious formats, such as those for exhibition panels and the pages of reports, brochures, newsletters, and so on. The grid functions as an invisible discipline in the communication of visual information. It also reduces costs, by standardizing sizes, eliminating typeface choices, and minimizing decision-making time.

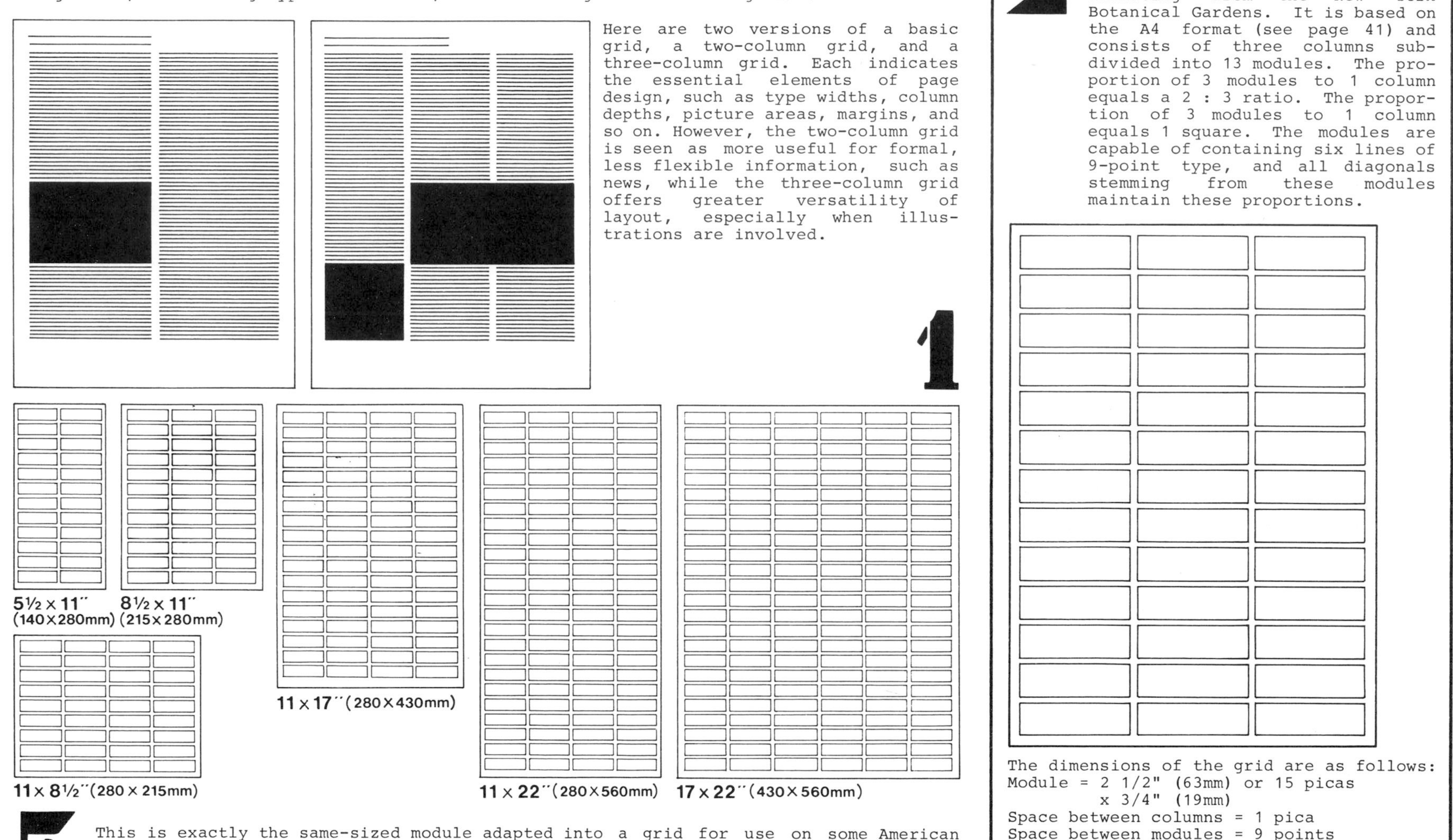

1

Here are two versions of a basic grid, a two-column grid, and a three-column grid. Each indicates the essential elements of page design, such as type widths, column depths, picture areas, margins, and so on. However, the two-column grid is seen as more useful for formal, less flexible information, such as news, while the three-column grid offers greater versatility of layout, especially when illustrations are involved.

5½ x 11" (140x280mm)

8½ x 11" (215x280mm)

11 x 8½" (280 x 215mm)

11 x 17" (280x430mm)

11 x 22" (280x560mm)

17 x 22" (430x560mm)

3

This is exactly the same-sized module adapted into a grid for use on some American paper sizes. In creating the 3 : 2 proportioned rectangle and the square, the modules create formats that are compatible with photographic prints taken with 2 1/4" and 35mm cameras for reproduction at a variety of scales without cropping.

2

This grid was designed by Vignelli Associates to coordinate the great variety of visual communications emanating from the New York Botanical Gardens. It is based on the A4 format (see page 41) and consists of three columns subdivided into 13 modules. The proportion of 3 modules to 1 column equals a 2 : 3 ratio. The proportion of 3 modules to 1 column equals 1 square. The modules are capable of containing six lines of 9-point type, and all diagonals stemming from these modules maintain these proportions.

The dimensions of the grid are as follows:
Module = 2 1/2" (63mm) or 15 picas
x 3/4" (19mm)
Space between columns = 1 pica
Space between modules = 9 points
Outside margin = 1/4" (6mm)
Inside margins = 3/16" (5mm)

The Anatomy of a Page Layout

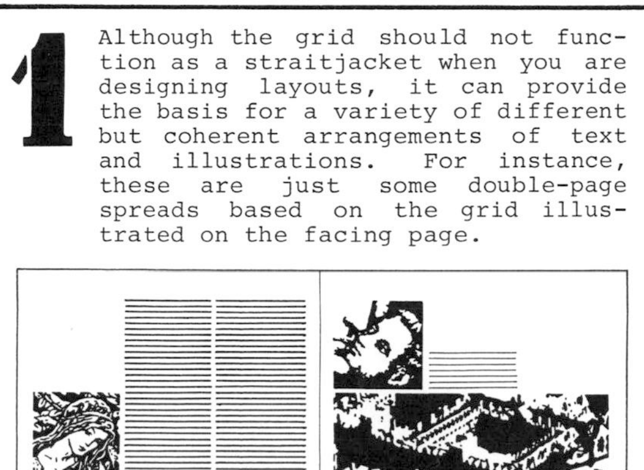

Although the grid should not function as a straitjacket when you are designing layouts, it can provide the basis for a variety of different but coherent arrangements of text and illustrations. For instance, these are just some double-page spreads based on the grid illustrated on the facing page.

This is another layout based on the same grid, together with a short glossary of technical terms used to describe the features of layouts.

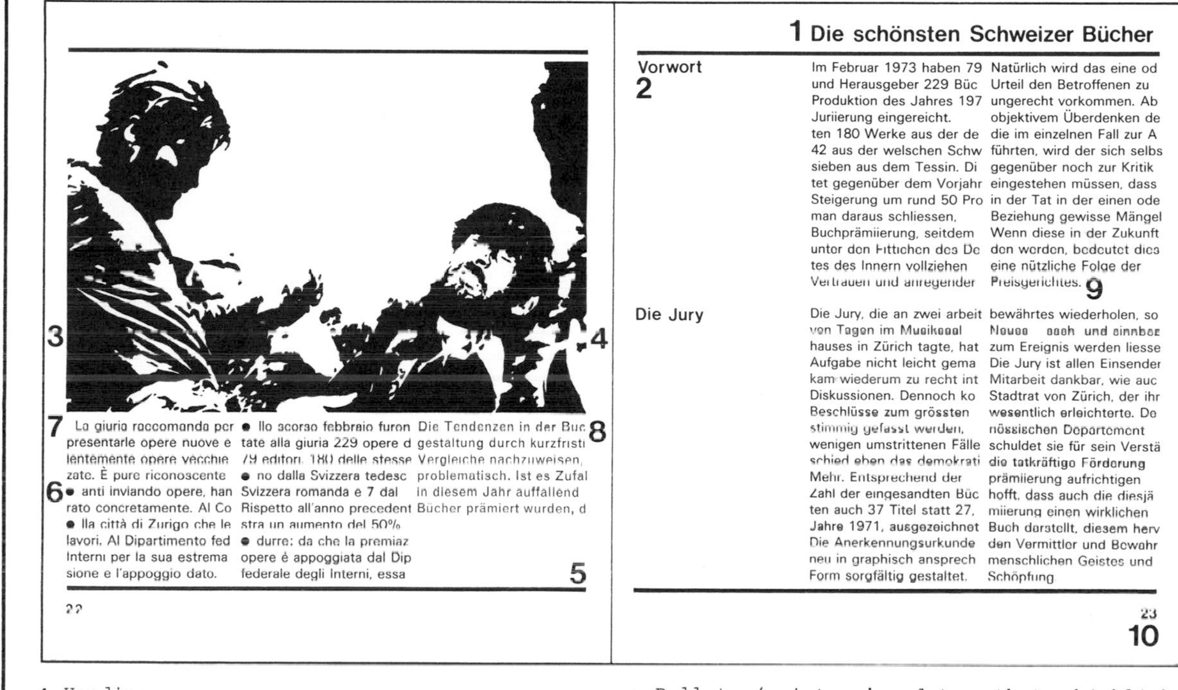

3 4

7 Lo giuria raccomanda per
presentarle opere nuove e
lentemente opere vecchie
zate. È pure riconoscente
6 ● anti inviando opere, han
rato concretamente. Al Co
● lla città di Zurigo che le
lavori. Al Dipartimento fed
Interni per la sua estrema
sione e l'appoggio dato.

● llo scorso febbraio furon
tate alla giuria 229 opere d
79 editori. 180 delle stesse
● no dalla Svizzera tedesc
Svizzera romanda e 7 dal
Rispetto all'anno precedent
stra un aumento del 50%
● durre: da che la premiaz
opere è appoggiata dal Dip
federale degli Interni, essa

Die Tendenzen in der Buc 8
gestaltung durch kurzfristi
Vergleiche nachzuweisen,
problematisch. Ist es Zufal
in diesem Jahr auffallend
Bücher prämiert wurden, d

5

22

1 Die schönsten Schweizer Bücher

Vorwort
2

Im Februar 1973 haben 79
und Herausgeber 229 Büc
Produktion des Jahres 197
Juriierung eingereicht.
ten 180 Werke aus der de
42 aus der welschen Schw
sieben aus dem Tessin. Di
tet gegenüber dem Vorjahr
Steigerung um rund 50 Pro
man daraus schliessen,
Buchprämiierung, seitdem
unter den Fittichen des De
tes des Innern vollziehen
Vertrauen und anregender

Natürlich wird das eine od
Urteil den Betroffenen zu
ungerecht vorkommen. Ab
objektivem Überdenken de
die im einzelnen Fall zur A
führten, wird der sich selbs
gegenüber noch zur Kritik
eingestehen müssen, dass
in der Tat in der einen ode
Beziehung gewisse Mängel
Wenn diese in der Zukunft
den werden, bedeutet dies
eine nützliche Folge der
Preisgerichtes. 9

Die Jury

Die Jury, die an zwei arbeit
von Tagen im Musiksaal
hauses in Zürich tagte, hat
Aufgabe nicht leicht gema
kam wiederum zu recht int
Diskussionen. Dennoch ko
Beschlüsse zum grössten
stimmig gefasst werden,
wenigen umstrittenen Fälle
schied eben das demokrati
Mehr. Entsprechend der
Zahl der eingesandten Büc
ten auch 37 Titel statt 27,
Jahre 1971, ausgezeichnet
Die Anerkennungsurkunde
neu in graphisch ansprech
Form sorgfältig gestaltet.

bewährtes wiederholen, so
Neues auch und sinnbar
zum Ereignis werden liesse
Die Jury ist allen Einsender
Mitarbeit dankbar, wie auc
Stadtrat von Zürich, der ihr
wesentlich erleichterte. De
nössischen Departement
schuldet sie für sein Verstä
die tatkräftige Förderung
prämiierung aufrichtigen
hofft, dass auch die diesjä
miierung einen wirklichen
Buch darstellt, diesem herv
den Vermittler und Bewahr
menschlichen Geistes und
Schöpfung

23
10

1 Heading

2 Subheads (act as introductions to new sections in a chapter)

3 Margin (the outer blank areas surrounding the printed matter)

4 Gutter (the inner margin of a double-page spread)

5 Rules (horizontal and vertical lines that, when used consistently, create distinctive visuals)

6 Bullets (printers' dots that highlight points in the text)

7 Indents (a typesetting less than a full column width, usually by an m space or two, that is, the square of any point size of type)

8 Caption (descriptive matter printed in conjunction with an illustration)

9 Widow (the last line of a paragraph, usually shorter than the column width)

10 Folio (page number)

N.B.: When an illustration is spread to the outer edge of a page, that is, without a margin, this is described as a "bleed."

How to Use Rules

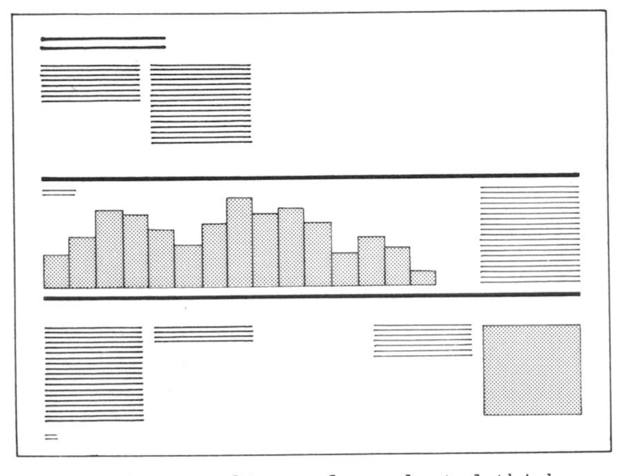

1

Rules are lines of a selected thickness that are used to structure layouts containing different kinds of information, a variety of different typefaces, or a mixture of type and illustrations, into a distinctive and unified whole.

2

Rules can also be used to distinguish one kind of information from another, such as captions from text, or--when faced with large areas of unrelieved text--can bring a contrasting element to break up the bland texture of a page.

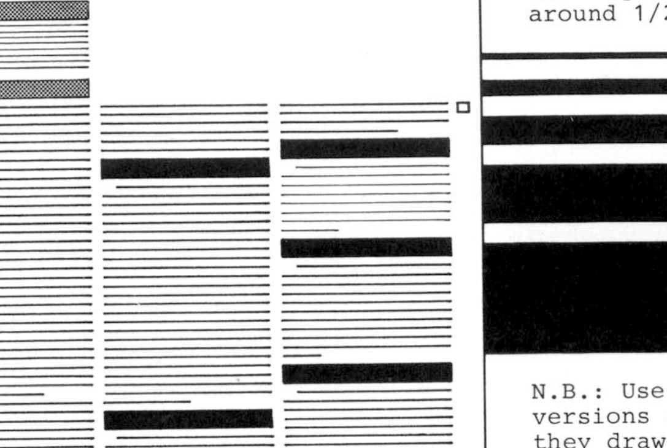

3

Rules can be ordered from your printer in a range of point sized widths. They are also available as dry-transfer border tapes dispensed from a tape pen and in dry-transfer run-on sheet form. These are easily applied to mechanicals and come in a range of widths from 1/64" (0.4mm), the equivalent of one point, to around 1/2" (12.7mm) in thickness.

N.B.: Use matte rules rather than the glossy versions and avoid fancy border tapes, since they draw attention to themselves rather than to the information they present.

4

When introducing rules into a mechanical, first make sure that the area to receive the tape is clean and free from any overlapping bits of pasted-up material. Then draw a straight guideline lightly in pencil.

N.B.: To ensure that a rule is straight when using the thinner lines on dry-transfer sheets, it is best to transfer them onto a strip of paper before gluing them into the mechanical.

Rules and Regulations

5

Next, draw just over the required length of tape from the dispenser and, holding this above the penciled guideline, make contact at each end by using the thumb at one end and the "pen" of the dispenser at the other.

N.B.: Rules can also be hand drawn into mechanicals using pen and ink. However, great care should be taken to achieve the precision offered by the dry-transfer equivalents.

Rules and Regulations

6

Press the tape into contact before cutting tape while lifting the dispenser toward the cut. Finally, trim away any excess tape at each end of the rule.

N.B.: Before rubbing down the tape, hold a ruler against its length to ensure that it is perfectly straight.

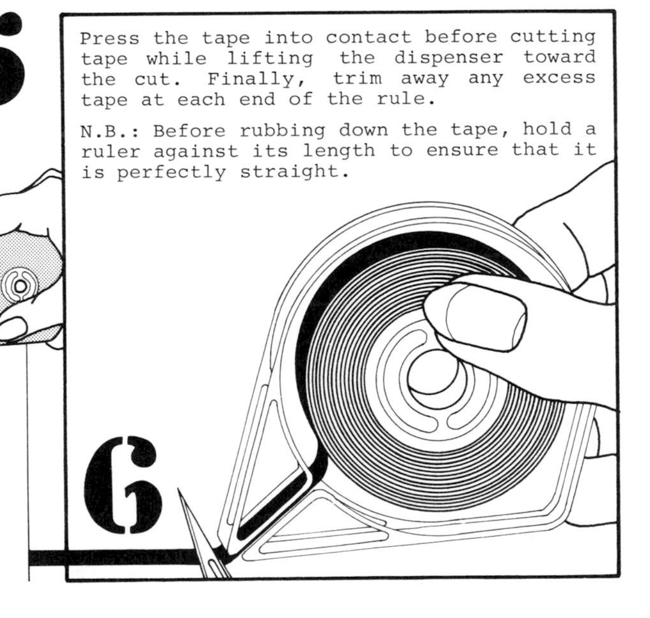

How to Use Dry-Transfer Lettering

Dry-transfer instant lettering makes an ideal medium for the production of headings, titles, subheads, and labels in mechanicals.

The size, number of lines, and extent of the lettering should first be established. Lettering for use on mechanicals can be worked on a separate sheet. This allows for some adjustment during its positioning in the layout.

1

Draw a guideline in nonrepro blue pencil on the paper and, after removing its backing sheet, place the carrier sheet face down so that the instant lettering guideline sits in position on the pencil guideline.

2

Work over the back of the carrier sheet with a brush handle or a burnisher until the letter appears gray. Make sure that all parts of the letter form are carefully transferred before removing the carrier sheet.

3

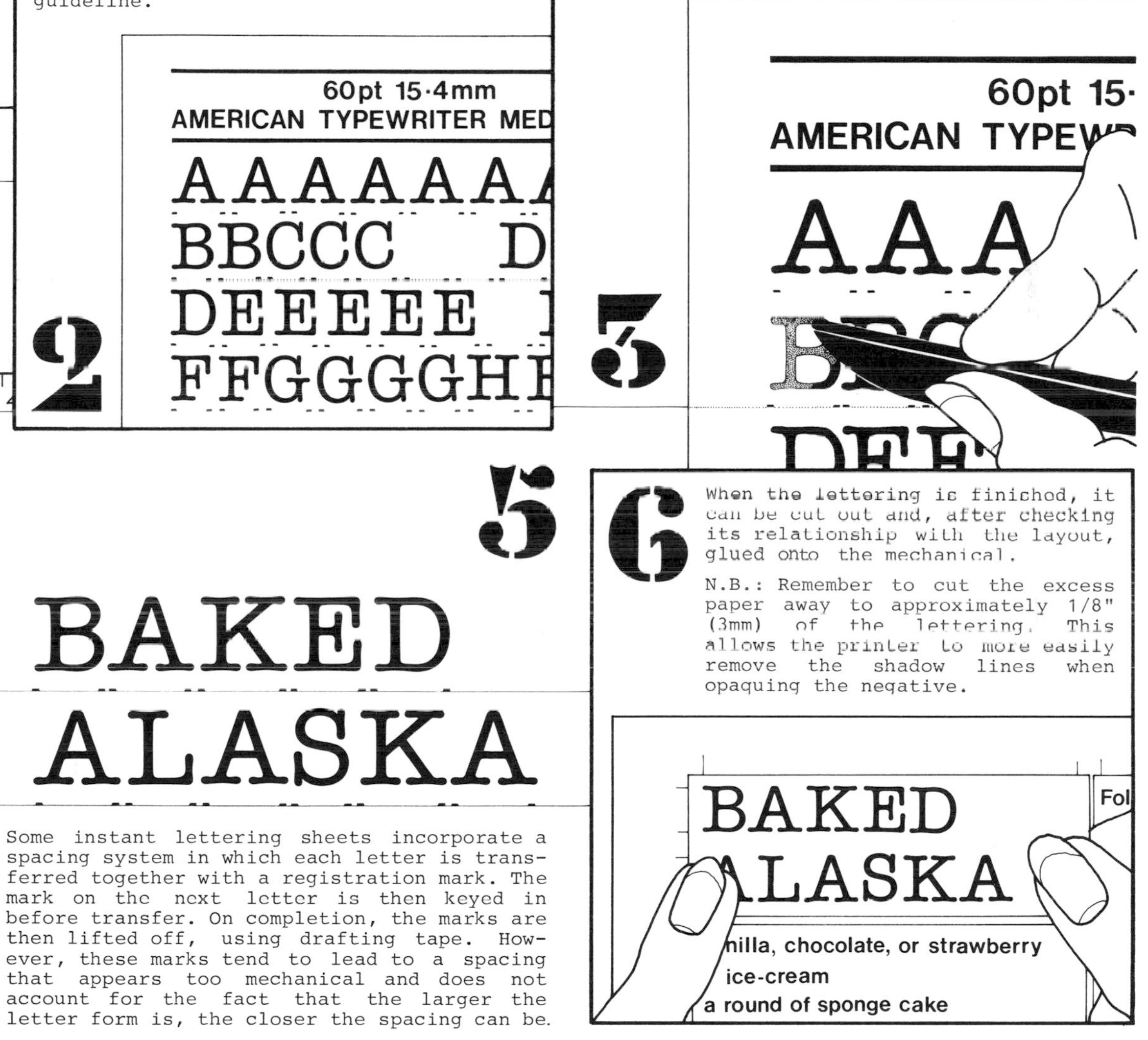

4

Once the letter forms are transferred, it is a good idea to burnish each letter in turn by covering it over with the backing sheet and rubbing with the brush handle or the burnisher.

N.B.: If a letter cracks, this can be filled in with a black pen and, if necessary, cleaned up with typewriter correction fluid. If a large chunk of letter becomes detached, this can be repaired with a pen or be corrected by transferring a replacement instant letter directly on top. Badly transferred letters can be plucked away using drafting tape.

5

Some instant lettering sheets incorporate a spacing system in which each letter is transferred together with a registration mark. The mark on the next letter is then keyed in before transfer. On completion, the marks are then lifted off, using drafting tape. However, these marks tend to lead to a spacing that appears too mechanical and does not account for the fact that the larger the letter form is, the closer the spacing can be.

6

When the lettering is finished, it can be cut out and, after checking its relationship with the layout, glued onto the mechanical.

N.B.: Remember to cut the excess paper away to approximately 1/8" (3mm) of the lettering. This allows the printer to more easily remove the shadow lines when opaquing the negative.

How to Produce a Mechanical (Paste-up)

1

There are two types of mechanical. One comprises pasted-up illustrations and text using galley proofs, for submission to the printer as a guide from which another version is produced. The second, and the one with which we are concerned here, comprises "camera-ready" artwork, that is, illustrations and text positioned as they will appear in the final film line shot after direct submission to the reproduction camera.

The paste-up process conforms to the grid layout (see page 52), against which the elements of a layout such as text, captions, images, and so on are arranged accurately in a consistent and balanced fashion.

2

As layouts are governed by the number of lines of type they contain, the grid should take account of the typeface size and leading. This can be included at the top of the grid as horizontal lines indicating the x-height, that is, the height of the letter "X" in the typeface used. Using a type scale, the number of lines required can be quickly measured.

N.B.: When using typewriters for text, type height and leading are fixed automatically.

3

A decision regarding the size of the mechanical has to be made. For example, will it include same-size copy, or will illustrations be worked separately and to a larger scale for later photographic reduction into the layout? A third option is to produce all camera copy to a size larger, such as half up, than that intended for publication. This decision relates directly to the required quality of print and to the level of complexity in illustrations.

This detail from a drawing has been reduced from an original size of 4 x 2 1/4" (100 x 56mm), that is, a 70 percent reduction.

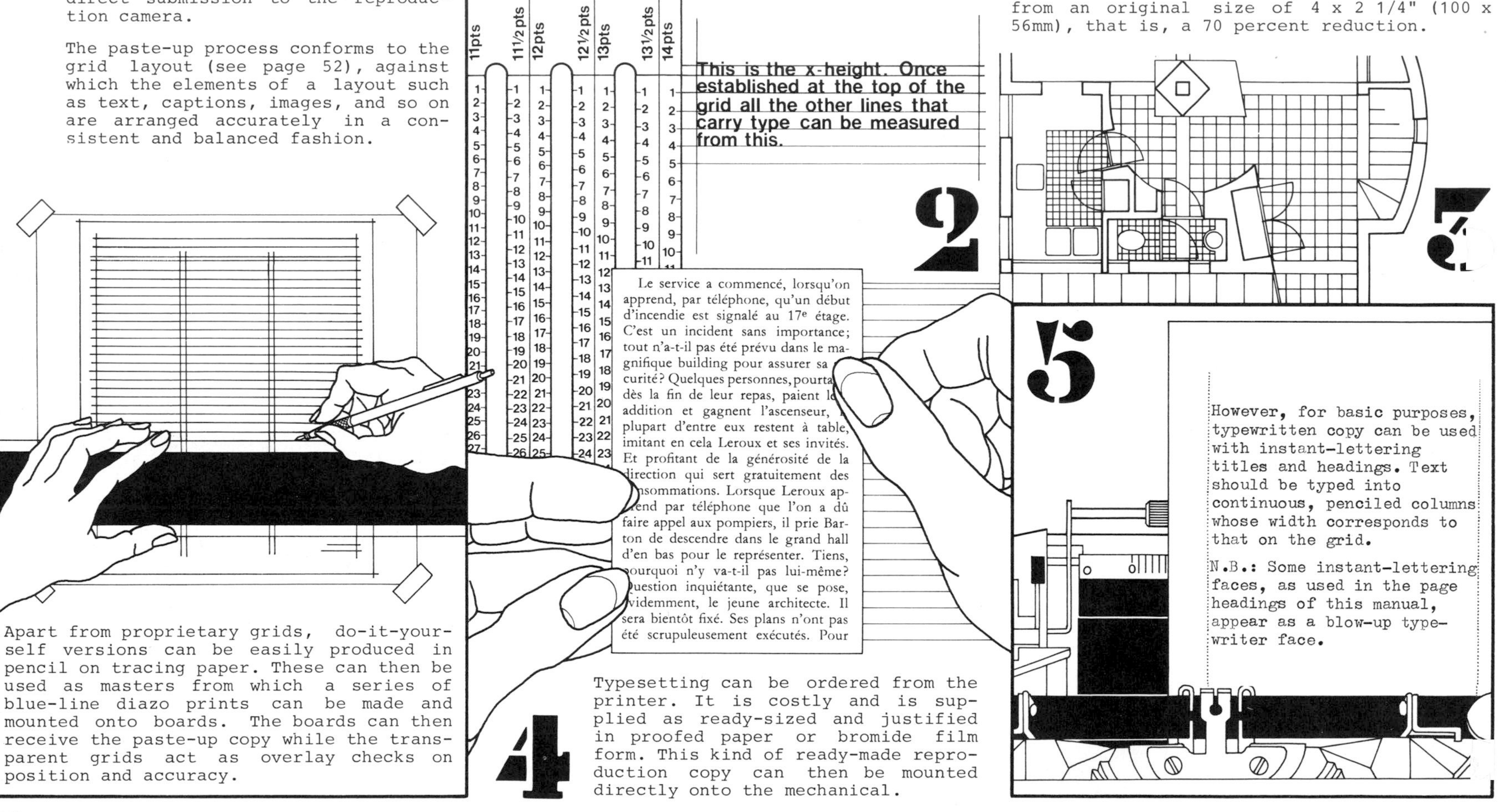

11 11½ 12 12½ 13 13½ 14
11pts 11½pts 12pts 12½pts 13pts 13½pts 14pts
This is the x-height. Once established at the top of the grid all the other lines that carry type can be measured from this.

Le service a commencé, lorsqu'on apprend, par téléphone, qu'un début d'incendie est signalé au 17e étage. C'est un incident sans importance; tout n'a-t-il pas été prévu dans le magnifique building pour assurer sa sécurité? Quelques personnes, pourtant, dès la fin de leur repas, paient l'addition et gagnent l'ascenseur, la plupart d'entre eux restent à table, imitant en cela Leroux et ses invités. Et profitant de la générosité de la direction qui sert gratuitement des consommations. Lorsque Leroux apprend par téléphone que l'on a dû faire appel aux pompiers, il prie Barton de descendre dans le grand hall d'en bas pour le représenter. Tiens, pourquoi n'y va-t-il pas lui-même? Question inquiétante, que se pose, évidemment, le jeune architecte. Il sera bientôt fixé. Ses plans n'ont pas été scrupuleusement exécutés. Pour

Apart from proprietary grids, do-it-yourself versions can be easily produced in pencil on tracing paper. These can then be used as masters from which a series of blue-line diazo prints can be made and mounted onto boards. The boards can then receive the paste-up copy while the transparent grids act as overlay checks on position and accuracy.

4

Typesetting can be ordered from the printer. It is costly and is supplied as ready-sized and justified in proofed paper or bromide film form. This kind of ready-made reproduction copy can then be mounted directly onto the mechanical.

5

However, for basic purposes, typewritten copy can be used with instant-lettering titles and headings. Text should be typed into continuous, penciled columns whose width corresponds to that on the grid.

N.B.: Some instant-lettering faces, as used in the page headings of this manual, appear as a blow-up typewriter face.

How to Produce a Mechanical (Paste-up)

6

Typeset or typewritten copy is cut into strips and, using a low-tack, aerosol adhesive or rubber cement, glued into position, making sure to align the lines of type with the baseline of the x-height lines on the grid.

N.B.: The adhesive should allow for the repositioning of paste-up material. Also, avoid touching the face of typewritten copy, as this will smudge easily.

7

If high quality reproduction is not vital, line drawings and even photographs can be glued directly onto the mechanical.

N.B.: The accurate positioning of copy material is crucial to a good visual effect. Therefore, when pasting up, use a T-square or a parallel-motion table to ensure that paste-ups are aligned.

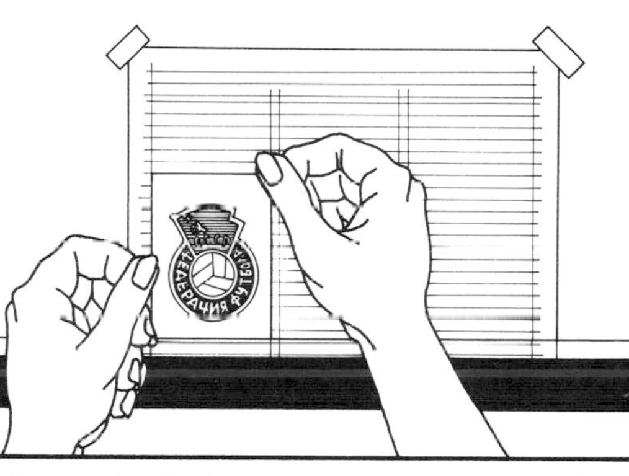

8

However, if halftone reproduction of an image is required, illustrations should be submitted to the printer independently of the mechanical. In this case the originals should have a transparent overlay attached to their faces. This sheet carries such information as reproduction dimensions, percentage of reduction, cropping requirements, and so on (see page 63).

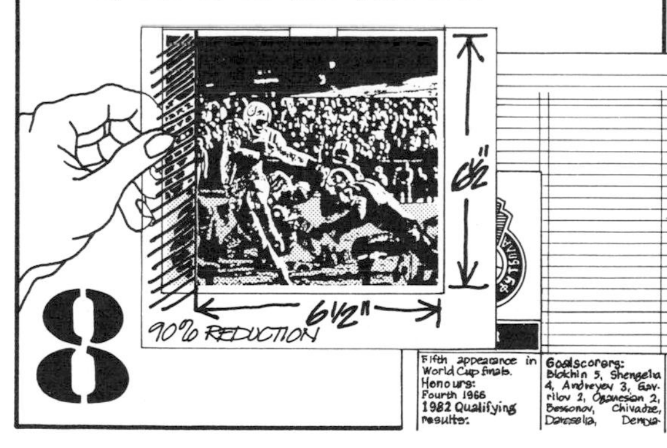

9

The mechanical, meanwhile, should carry outside corner marks drawn in a fine black line to show the precise location of each halftone illustration.

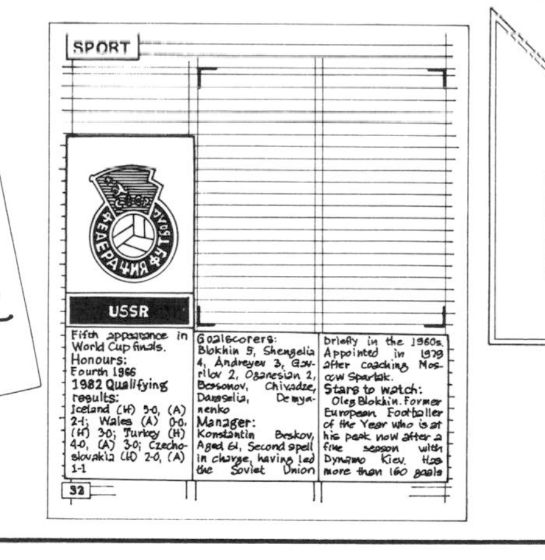

10

The halftone originals will be photographed separately through a cross-hatched screen. Meanwhile, their exact areas on the layout mechanical will be blanked off so that, when this is photographed, they appear as clear windows on the plastic film negative. The halftoned and sized negatives are then pasted onto the mechanical negative, this integration providing the final film ready for opaquing, that is, painting out any blemishes, and printing.

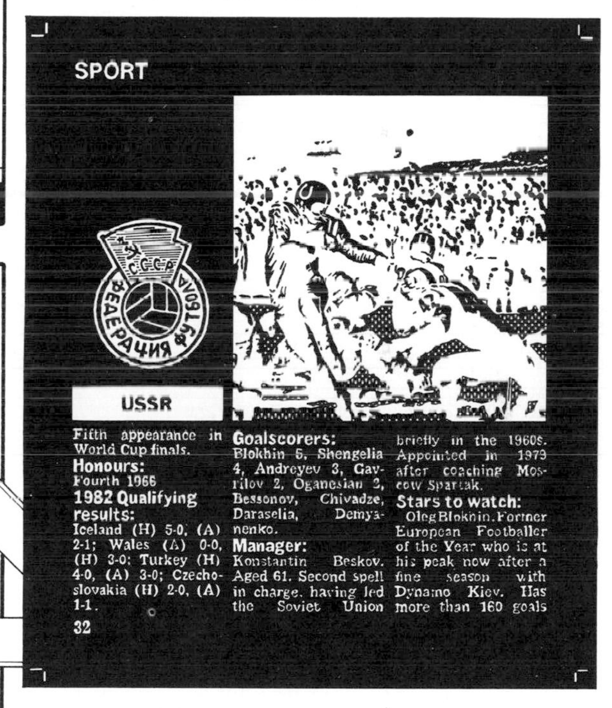

N.B.: Halftone reproduction varies in process and cost. For example, photoengraving is high quality but expensive. The Photo Mechanical Transfer (P.M.T.) machine, though it provides coarser negatives, is cheaper and can achieve a wide range of screened effects.

Some Hints when Making Paste-ups

It is worth mentioning that shadows cast from the edges of thicker papers mounted on layouts will tend to print. Excessive buildup of overlaid materials should thus be avoided.

For this reason, columns of typed or typeset text should be trimmed into columns with a margin of about 1/8" (3mm) away from each side edge of the type. This margin allows the printer to easily eradicate them on the film negative during the opaquing stage.

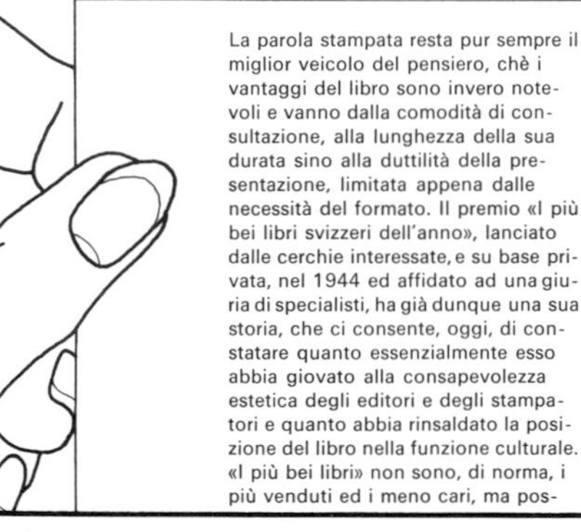

La parola stampata resta pur sempre il miglior veicolo del pensiero, chè i vantaggi del libro sono invero notevoli e vanno dalla comodità di consultazione, alla lunghezza della sua durata sino alla duttilità della presentazione, limitata appena dalle necessità del formato. Il premio «I più bei libri svizzeri dell'anno», lanciato dalle cerchie interessate, e su base privata, nel 1944 ed affidato ad una giuria di specialisti, ha già dunque una sua storia, che ci consente, oggi, di constatare quanto essenzialmente esso abbia giovato alla consapevolezza estetica degli editori e degli stampatori e quanto abbia rinsaldato la posizione del libro nella funzione culturale. «I più bei libri» non sono, di norma, i più venduti ed i meno cari, ma pos-

1

2

Prior to paste-up, it is also worthwhile to check the entire layout as a dry run. If satisfactory, begin to paste up text and illustrations, working from the bottom of the grid upward. This ensures a clean and accurate baseline.

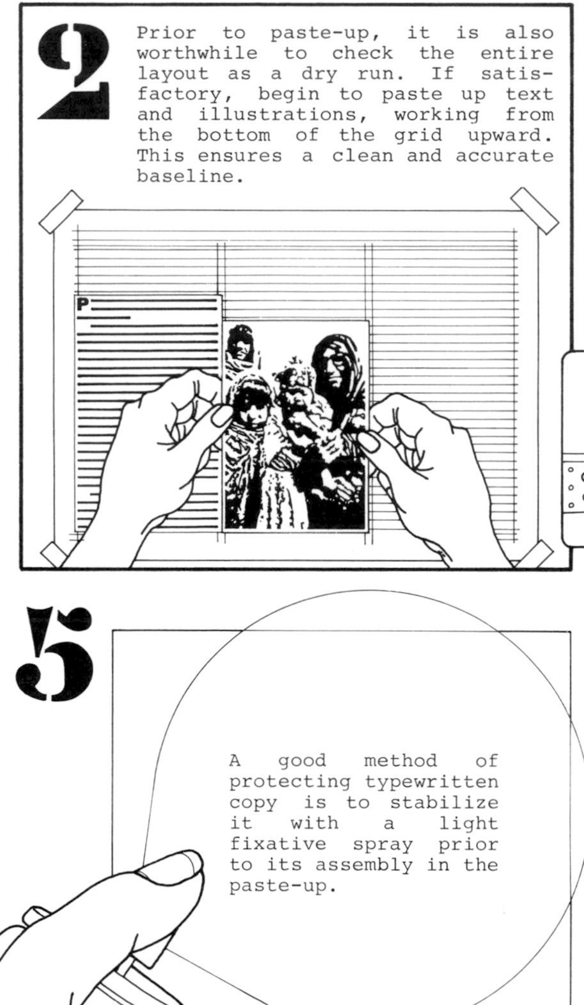

3

Remember to continue checking carefully the horizontal alignment of lines of type. A constant check should also be made on the vertical alignment of columns. In order to maintain a disciplined layout, the edges of illustrations and graphics should line up with the edges of columns.

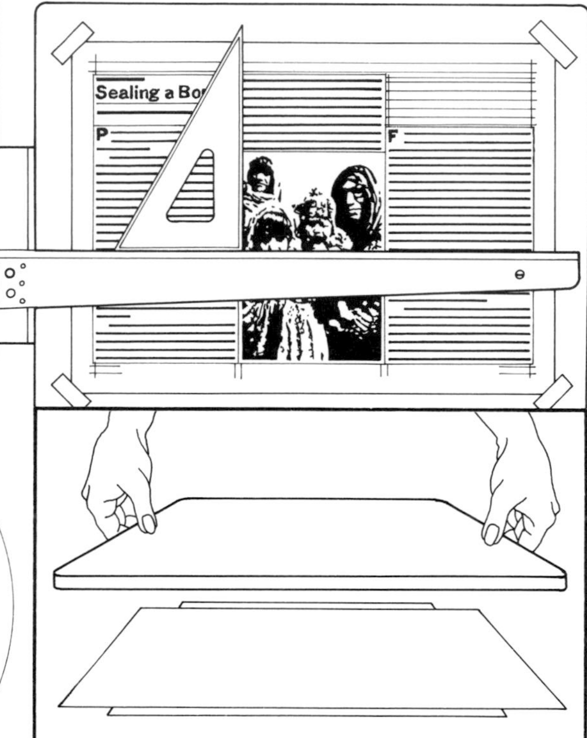

4

Any blemishes or dirty marks on the completed layout can be removed using a soft eraser, rubber-cement pickup, or typist's correction fluid. However, it is important not to attempt any vigorous cleaning of typewritten copy because, as previously mentioned, it is susceptible to smearing.

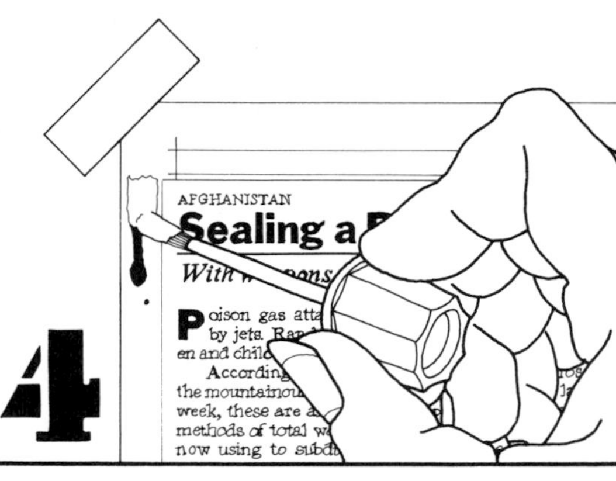

5

A good method of protecting typewritten copy is to stabilize it with a light fixative spray prior to its assembly in the paste-up.

6

Finally, protect the completed mechanical with a sheet of clean paper and, to ensure that all pasted-up material remains flat, place it face up under the weight of a drawing board.

It is always a good practice to make a last-minute proofreading and layout check of the completed mechanical.

How to Strip In Corrections

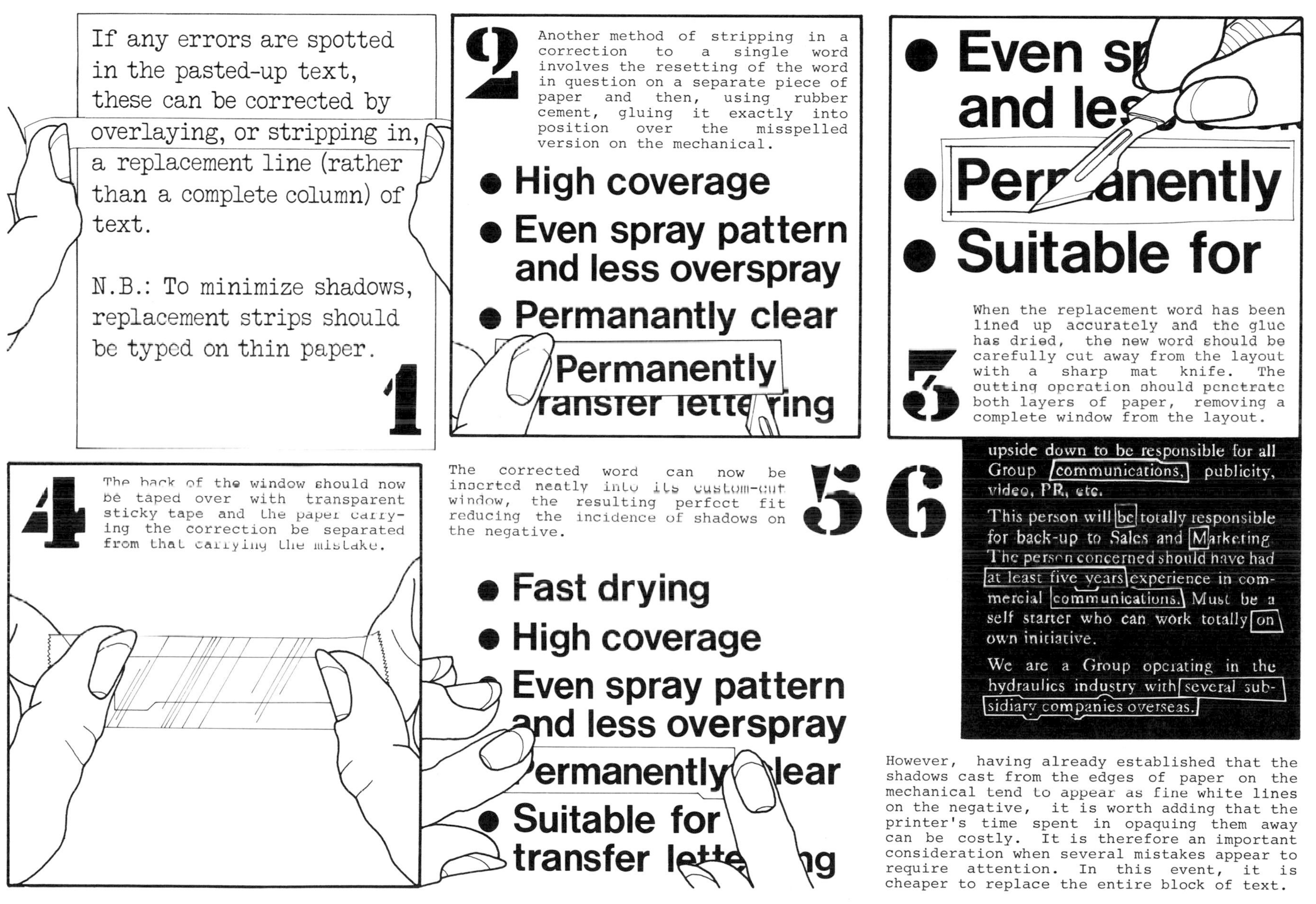

• High coverage
• Even spray pattern and less overspray
• Permanantly clear
Permanently
ransfer lette ring
• Even sp
and les
• Permanently
• Suitable for
upside down to be responsible for all Group communications, publicity, video, PR, etc.
This person will be totally responsible for back-up to Sales and Marketing. The person concerned should have had at least five years experience in commercial communications. Must be a self starter who can work totally on own initiative.
We are a Group operating in the hydraulics industry with several subsidiary companies overseas.
• Fast drying
• High coverage
• Even spray pattern and less overspray
Permanently clear
• Suitable for transfer lettering

1 If any errors are spotted in the pasted-up text, these can be corrected by overlaying, or stripping in, a replacement line (rather than a complete column) of text.

N.B.: To minimize shadows, replacement strips should be typed on thin paper.

2 Another method of stripping in a correction to a single word involves the resetting of the word in question on a separate piece of paper and then, using rubber cement, gluing it exactly into position over the misspelled version on the mechanical.

3 When the replacement word has been lined up accurately and the glue has dried, the new word should be carefully cut away from the layout with a sharp mat knife. The cutting operation should penetrate both layers of paper, removing a complete window from the layout.

4 The back of the window should now be taped over with transparent sticky tape and the paper carrying the correction be separated from that carrying the mistake.

5 The corrected word can now be inserted neatly into its custom-cut window, the resulting perfect fit reducing the incidence of shadows on the negative.

6 However, having already established that the shadows cast from the edges of paper on the mechanical tend to appear as fine white lines on the negative, it is worth adding that the printer's time spent in opaquing them away can be costly. It is therefore an important consideration when several mistakes appear to require attention. In this event, it is cheaper to replace the entire block of text.

Sizing and Scaling Illustrations

1

When artwork is to be reduced for insertion into a layout, a simple method of sizing the relative formats is employed.

First, tape a tracing-paper overlay over the face of the original.

2

Using a pencil and a ruler, trace the frame of the picture area.

Then, using a blue nonprinting pencil, establish the reproduction on the layout by drawing its frame, together with a diagonal line that connects the bottom lower-left corner with the top right corner.

4

Next, superimpose the tracing-paper overlay on the layout format, registering the two frames at their lower left corners. Extend the diagonal line until it connects with the frame of the original.

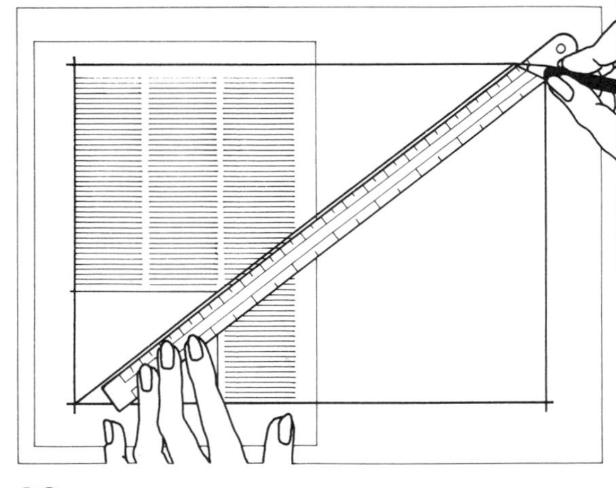

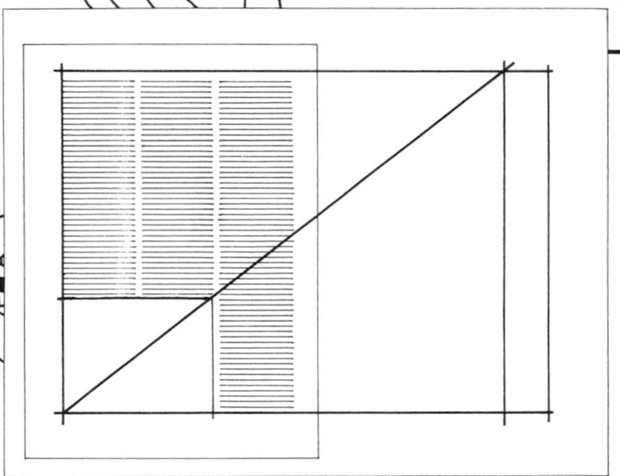

5

Draw a line from this intersection parallel to the adjacent side of the frame. This now establishes the exact proportional relationship of the original so that, after reduction, it fills the area reserved for it in the layout.

6

Replace the overlay over the original and use the adjusted frame as a window through which to select the best picture area for reproduction. Once this is determined, tape the overlay into position.

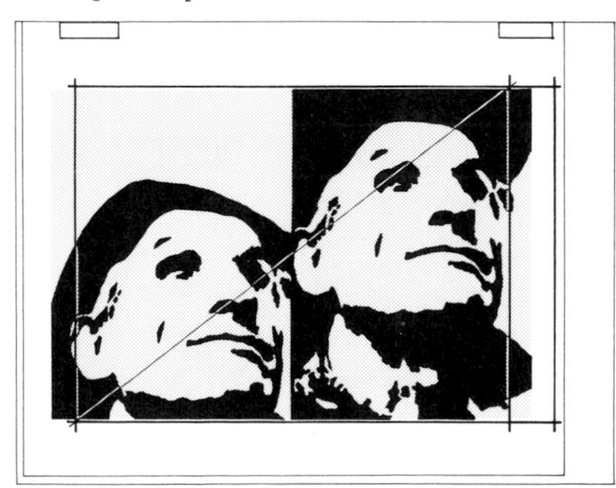

Sizing and Scaling Illustrations

7

To complete this stage, the areas of the image to be cropped (that is, those areas not included in the reproduced image), should be shaded and the dimensions of the reproduction size indicated.

8

Another piece of information that should be included on the overlay is the percentage of reduction. This is important because most reproduction cameras are calibrated by percentage rather than size. The percentage of reduction or enlargement can be quickly calculated by using a proportion wheel (a reproduction computer) as described next.

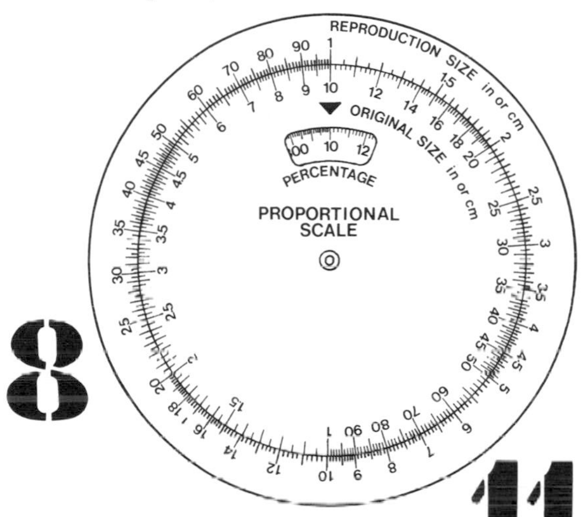

9

Having already established the dimensions of the original in relation to its intended reproduction size, it is best to begin calculating with the least flexible dimension in layouts, that is, the width.

Find the width dimension of the original on the inside wheel of the calculator, then the width dimension of the reproduction format on the outer wheel. Rotate the calculator until the two match up.

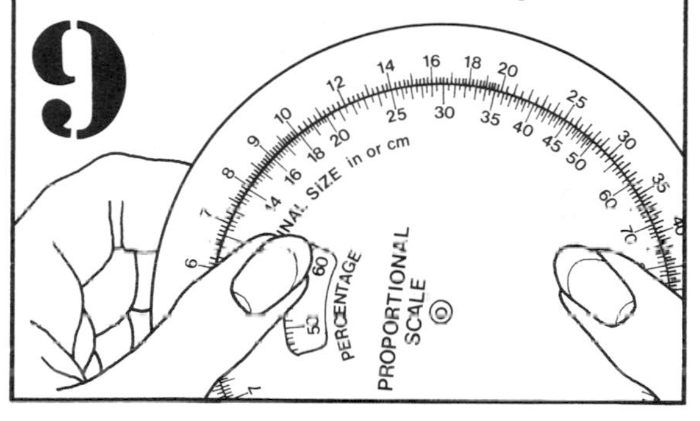

10

The percentage for that reduction ratio will now be indicated by the arrow in the small window.

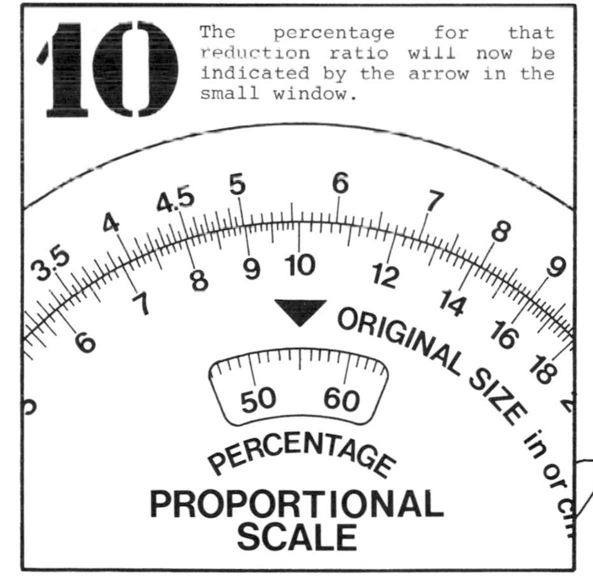

11

To check the relative heights of the image at that percentage, do not disturb the wheel. Simply find the height of the original on the inside wheel. The corresponding height on the outer wheel will give the height of its reproduction size at that percentage.

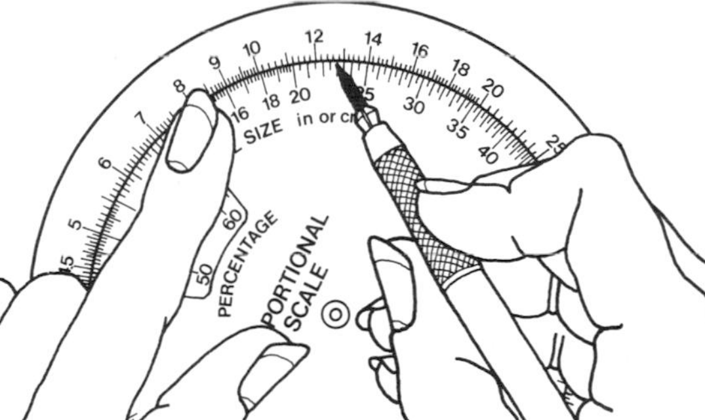

12

Alternatively, a pocket calculator can be used to work out the reduction percentage. Simply divide the width of the reproduction size by the width of the original and multiply this by 100. To check the reproduction height, just multiply this percentage number by the height of the original and divide by 100.

$$\frac{\text{Reproduction Width} \times 100}{\text{Original Width}} = \text{Percentage Reduction}$$

$$\frac{\text{Percentage Reduction} \times \text{Original Height}}{100} = \text{Reproduction Height}$$

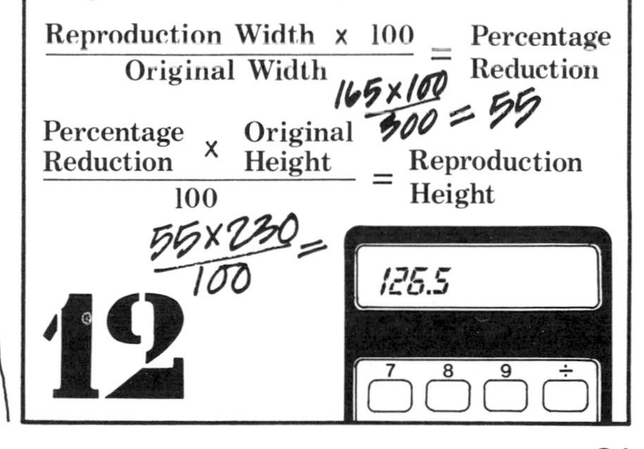

Hints when Submitting Photographs for Publication

1

Always specify a glossy finish when ordering monochrome prints for reproduction. They should also be submitted unmounted. Crop marks inform the printer which area of the print is required for reproduction. If not indicated on an overlay, crop marks should be placed on the margins of the print using a nonsoluble ink, as this avoids smudging the surface.

N.B.: Avoid handling the face of a print, and never attach it to other material with paper clips.

2

Color reproduction is rarely made from color prints. However, if this is the only form of material available, avoid a silk finish. This finish is embossed into the surface after the print is made and makes for difficult reproduction.

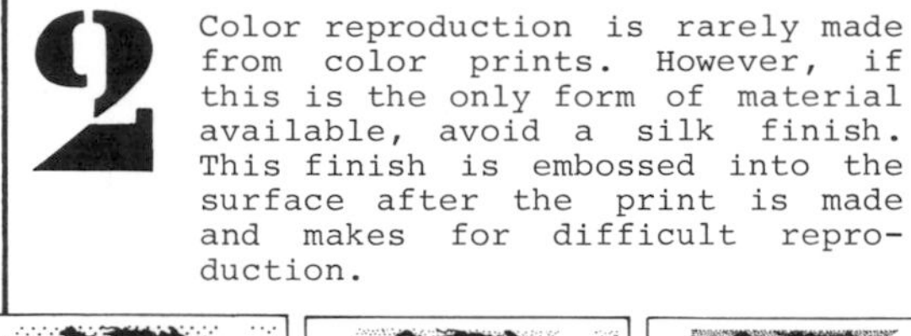

N.B.: Avoid making big enlargements from prints, and never use Polaroid prints.

3

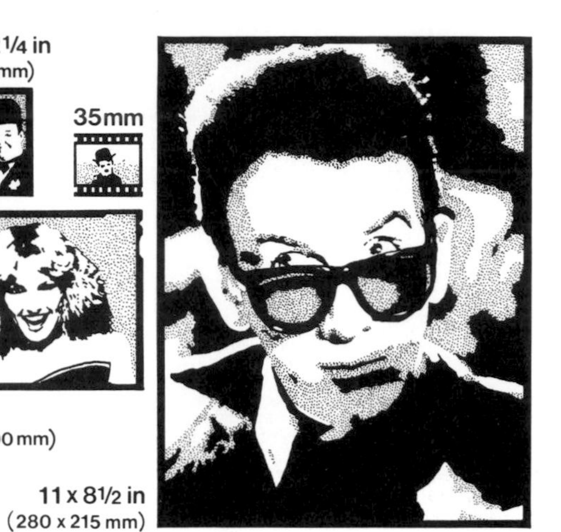

For color reproduction, most printers insist upon transparencies. As a general rule, the bigger the transparency, the better the quality of reproduction.

4

Information detailing the reproduction requirements for a transparency should be marked on a transparency overlay. This could carry the outline of the required image, its dimensions, and any cropping.

N.B.: When submitting 35mm slides, remember that the transparency is usually removed from its mount.

5

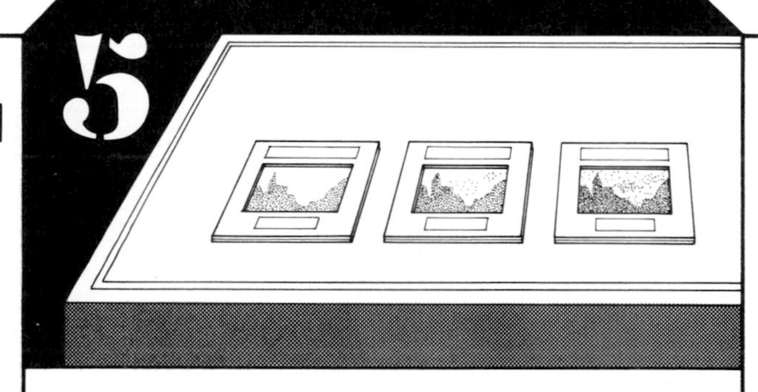

When shooting other than action situations, try to bracket your shots: make one shot at what you believe to be the correct aperture, then another with the lens closed down a half stop and a third with the lens opened a half stop. Apart from the fact that bracketing offers a choice of negatives, colors appear richer in underexposed transparencies.

N.B.: The lower the ASA film number, the less grainy the reproduction. Thus a film rated at ASA 64 will tend to give sharper reproduction than one rated at ASA 200.

6

Special filters are available to avoid green color casts produced when shooting under fluorescent lights. There is one for warm fluorescent lamps and another for cool fluorescent lamps. When shooting outdoors on sunny days, better results will be obtained by using a polarizing filter.

N.B.: When using hand-held exposure meters, run exposure tests. The filter factor for most polarizers is 3X or 4X, despite what the data sheet may say.

Special Effects with Overlays

1

Line drawings or typefaces can be integrated into a photograph original to create exciting composite pictures. However, the additional graphics have to be worked on the face of a transparent overlay because, if applied directly to the photograph, lines would lose their contrast value, while press-on lettering would reveal its imperfections and edge shadows to the continuous tone of the halftone camera's voracious eye.

The overlay is placed over the photograph, is masking taped to the top edge, and the additional graphics are worked in solid black ink or another photo opaque medium. So that legibility is achieved in the composite picture, black lines and type should coincide with light areas on the print, while white lines and type should coincide with dark areas of the print.

N.B.: The printer makes two films from the artwork. For white lines and type a film positive is made from the overlay, a negative being made from the print. For black lines and type, negatives are made from both overlay and print.

Before being submitted for process photography and printing, the image on the overlay and that of the photograph must be carefully registered. To do this, apply three or more registration marks outside the area of the picture on both print and overlay. These will ensure a perfect match of the two when integrated.

2

3

Another use for overlays is when a drawing is too large to comfortably receive dry-transfer screens. In this case the screens are applied at the printer's after reduction. In order to communicate your requirements, supply a separate overlay for each of the different tints required.

Each overlay is fixed in turn on the drawing, and the precise area of each tint is traced and filled in with solid black ink or paint.

N.B.: It is important to remember that one overlay represents one screen tint, the number of overlays reflecting the number of tints intended for the drawing.

Each overlay should also carry the required density and grade of tint together with registration marks traced from those initially drawn outside the picture area on the drawing itself.

Hints when Submitting Copy for Publication

1

When preparing copy for typesetting, type the text on one side only of the paper. Set the typewriter to type double or triple linespacing so that there is a clear space between lines for instructions and corrections. Also, leave plenty of margin to the left and right of the column and number each page clearly in the top right-hand corner.

N.B.: Make sure that the copy is correctly spelled and punctuated. Also, type illustration captions on a separate sheet, making sure that they are annotated clearly with their textual references.

(10)
When preparing copy for typesetting,
type the text on one side only of the
paper. Set the typewriter to type
double or triple linespacing so that
there is a clear space between lines
for instructions

2

When ordering typeface, its size and font (style of type) should be clearly specified on the copy, together with any special instructions, such as words, subheads, and so on, that need to be capitalized, italicized, set in boldface, or in a different size.

8/10 H.M. x 18 p.i.
FL & FR

N.B.: A translation of the above abbreviated instruction is as follows. Set in 8 point type with 10 point line-spacing. The type should be set in Helvetica Medium and the columns be 18 picas wide. The columns should be flush left and flush right.

Copy for a book appears first in type at the galley proof stage, in which the text appears in one continuous column for checking. A later proofing stage occurs when the corrected galley is converted into page form, including proofs of any illustrations. These may appear as photographic prints known as bromides or blues or, if in color, as scatter proofs. The proof stages represent filters that occur after the corrections to the original manuscript. Any mistakes that may be overlooked during any of these stages will find their way into the printed result. It is crucial therefore that proofs be scrutinized meticulously. It is also important that your means of communicating corrections and instructions to the typesetter conform to the internationally standardized language of proof correction, some of which is shown below.

N.B.: As the typesetter will not read the entire text but will look only for corrections indicated on the manuscript or proofs, the reader must record corrections clearly using two signs, one in the text and one in the margin.

TYPOGRAPHICAL INSTRUCTIONS

SIGN IN TEXT	SIGN IN MARGIN	MEANING
proofreaders' signs	(cap)/	Capitalize lowercase letter
proofreaders' signs	(lc)//	Change to lowercase
proofreaders' signs	(sc)//	Set in small capitals
proofreaders' signs	(BF)//	Set in boldface type
proofreaders' signs	(ital)//	Set in italic type
proofreaders' signs	(rom)//	Set in roman type
proofreaders' signs	(WF)/	Wrong font; set in correct type
proofreaders' signs	(X)/	Reset broken letter
proofreaders' signs	9/	Reverse (type upside down)

PUNCTUATION INSTRUCTIONS

SIGN IN TEXT	SIGN IN MARGIN	MEANING
proofreaders' signs	,/	Insert comma
proofreaders' signs	' /	Insert apostrophe or single quotation mark
proofreaders' signs	" / "	Insert quotation marks
proofreaders' signs	⊙/	Insert period
proofreaders' signs	?/	Insert question mark
proofreaders' signs	;/	Insert semicolon
proofreaders' signs	:/	Insert colon
proofreaders' signs	=/	Insert hyphen
proofreaders' signs	M/	Insert em dash
proofreaders' signs	N/	Insert en dash

OPERATIONAL INSTRUCTIONS

SIGN IN TEXT	SIGN IN MARGIN	MEANING
proofreaders' signss	₰	Delete
proofreaders' s igns	⁀	Close up; delete space
prooofreaders' signs	₰	Delete and close up
proofreaders/signs	(#)	Insert space
some/proofreaders/signs	(eq#)	Make space between words equal
signs proofreaders'	(tr)	Transpose
proofreaders' signs	(stet)	Leave as printed
signs	proofreaders'/	Insert new matter
]proofreaders' signs	]/	Move to the right
proofreaders' signs[	[/	Move to the left
□proofreaders' signs	□/	Indent 1 em (width of letter M)
]proofreaders' signs[	][/	Center over column
(3)proofreaders' signs	(SP)/	Spell out
3pt# proofreaders' signs	(+#)/	Add this much space between lines
prfreaders' signs	o/	Insert missing letter
proofreaders' signs	(align)/	Straighten type; align horizontally
proofreaders' signs	‖/	align vertically
proofreaders' signs. Many	¶/	Begin new paragraph here
proofreaders' signs. Many	(no ¶)/	Run on (no paragraph here)

3 BASIC PRINTMAKING PROCESSES

Basic Printing Techniques

1

Relief printing (block printing) involves the transfer under pressure of ink, paint, or dye from a raised surface. This surface can be fashioned using special gouging tools, such as in lino or woodcuts, so that the unwanted, or negative, elements of an image are cut away. Alternatively, when a textural effect rather than a precise design is required, prints can be made from any surface that is elaborate in relief, such as that on leaves, bark, vegetables, and fruit.

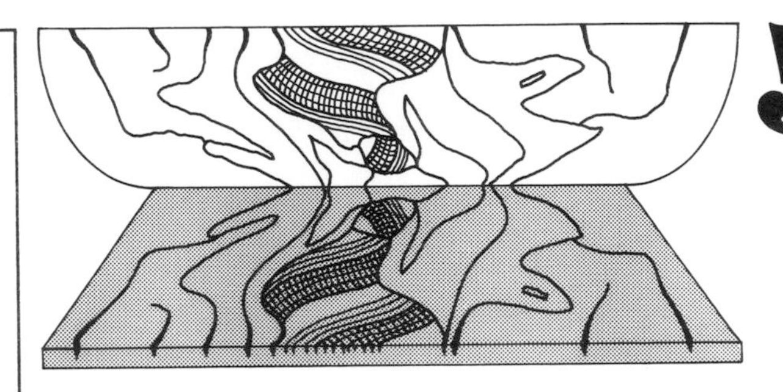

2

By contrast, intaglio prints are made by pressing paper into inked indentations on a metal surface. In this technique the positive elements of a design are incised to leave raised surface plateaus as the negative elements. As its name implies, etching relies on an acid to bite into the positive areas of the plate, whereas the engraving and drypoint techniques employ special tools to cut or score the design.

5

Artists such as Rembrandt, Vuillard, and Picasso have demonstrated that such printmaking techniques as etching, lithography, and linocutting can be used as mediums for spontaneous sketching and direct drawing. However, it is more common to transfer by tracing an established design into a sympathetic mode of printing. When color is involved, this transfer process necessitates the analytical breakdown of an image into its constituent hues, each block, plate, or screen representing the required areas of an individual color--both for reproducing it and for color mixing it when it is overprinted with others in the printing sequence.

3

Stencil printing involves the adhesion or enmeshing of a paper, wax, plastic, or liquid mask to a stretched fabric screen. In screen-printing the stencil masks off the negative areas of a design while ink is forced through the exposed, positive areas of the mesh and onto the print paper.

4

Planographic printing, on the other hand, describes the taking of an inked image from a flat surface. In using this technique, lithography relies upon the antipathetic nature of grease and water, and involves the drawing of a positive image with a greasy crayon or liquid on a prepared metal or stone surface. During printing the grease attracts and receives the printing ink, while by repelling it water provides the negative image. The technique of marbling, which lifts oil-based inks from the surface of water, can also be described as planographic.

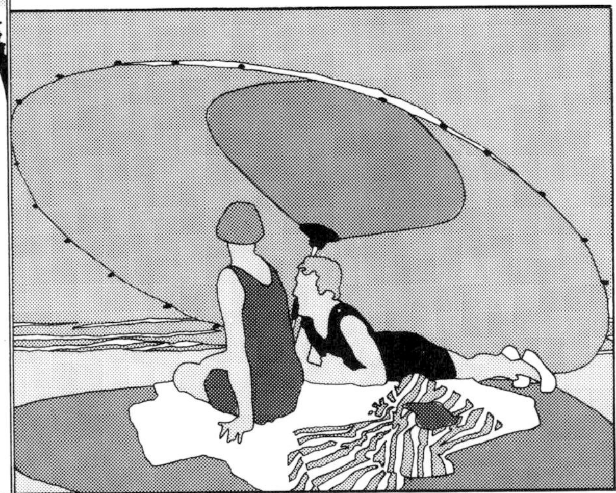

Experimental Relief Prints and Monoprints

1 Monoprinting is the making of individual prints. Basically, it involves rolling up a glass plate with block-printing ink, covering this with thin paper, and, using a pencil for line and the thumb for value, drawing over the back of the sheet in order to effect a transferred image.

N.B.: A negative of the same image can be gained by hand burnishing a fresh sheet of paper placed over the glass plate.

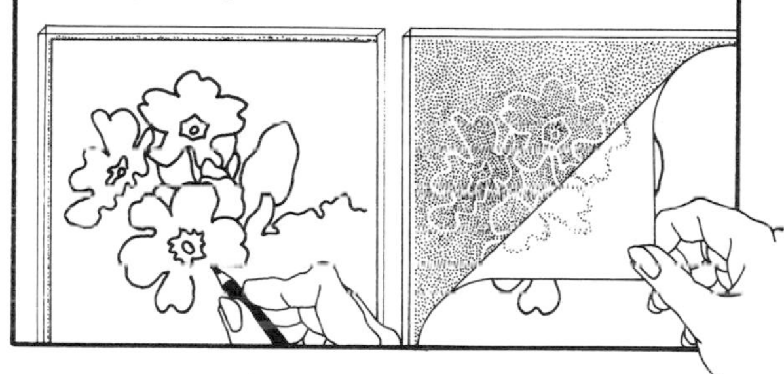

2 There are many methods of achieving monoprints. The one that offers the most control is the positioning of cut-out paper shapes inked up in one color, on a glass plate rolled up in another color. Once they are covered with thin paper, impressions can be taken by passing the plate through a mangle with soft rollers, or by burnishing with a clean hand roller.

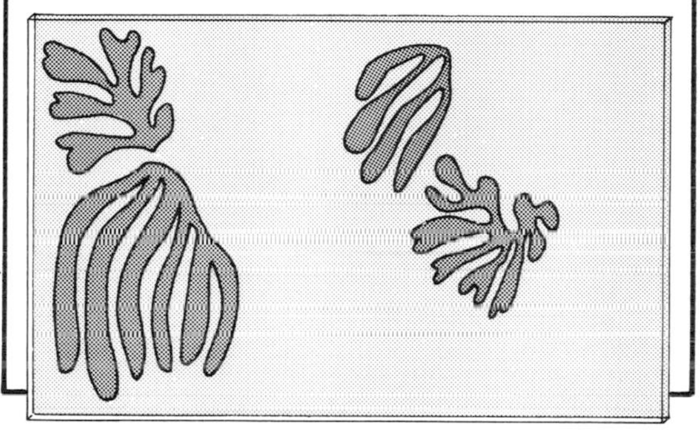

3 More complicated images can be achieved by positioning a variety of differently ink-colored paper stencils on an ink-rolled glass plate. After they are printed, this setup can then be further developed by altering the positions of the stencils before reprinting, or by reintroducing the proof for overprinting.

4 The simplest method of making a relief print is via the potato cut. A potato is sliced, then a raised shape carved on one face to receive a painted coat of watercolor or poster color before making its impression on paper.

N.B.: Effective patterns can be constructed by repeating the potato print regularly. These can exploit both a rotation of the "block" and the potential of impressions at various intensities.

5

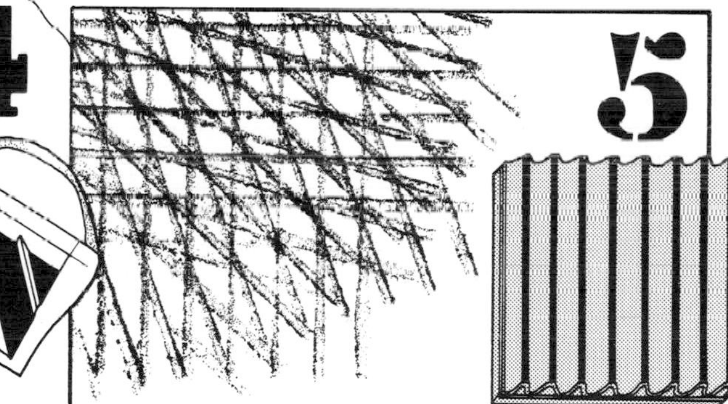

Block prints can be made from any suitable natural or manmade material, such as hessian, string, wood grains, corrugated cardboard, and many other substances. The selected surface is glued to a thin baseboard and, when dry, inked up with a roller for impressions made either by pressure from the back of the block or by hand burnishing the back of the paper.

6 Interesting effects can also result from rolling a well-charged ink roller once over a flat piece of material, such as net curtain, and transferring its impression to a smooth piece of paper. Although the transferred image is a negative, this indirect-printing method represents the principle of offset reprography.

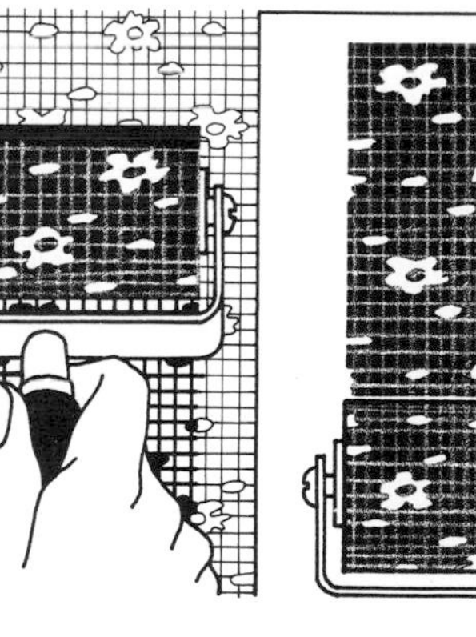

How to Make a Linocut

1 The linocut is the most common means of producing hand-cut relief-printing blocks, linoleum being a softer medium for gouging than wood or metal.

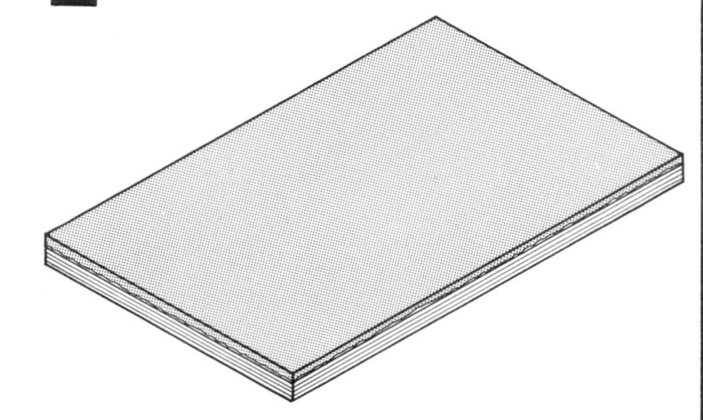

It is a good idea to glue thinner lino sheets onto a same-sized plywood backing block to stabilize their surfaces during both the gouging and the proofing stages.

A good tip is first to coat the smooth face of the lino with white or a light-colored water-based paint. When dry this provides a contrasting background for the subsequently pencil-transferred artwork at the gouging stage, and a clearer definition between the positive raised plateaus of lino that will print and the negative areas cut away.

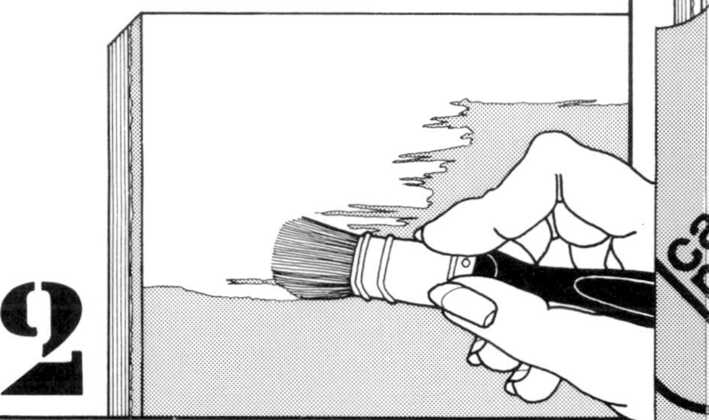

2

3 After the guidelines required for the transfer have been pencil traced from the master design, the tracing paper is reversed laterally, so that it becomes a mirror image of the original, and overlaid on carbon paper before retracing onto the lino block.

Lines and the outlines of positive shapes should be gouged first, using a fine V-shaped gouge or a scalpel blade. Once a linear framework is established, the broader negative areas are then removed by deep cutting with a U-shaped gouge.

N.B.: Gouges should always be worked away from the body, away from the other hand, and away from established fine lines and outlines.

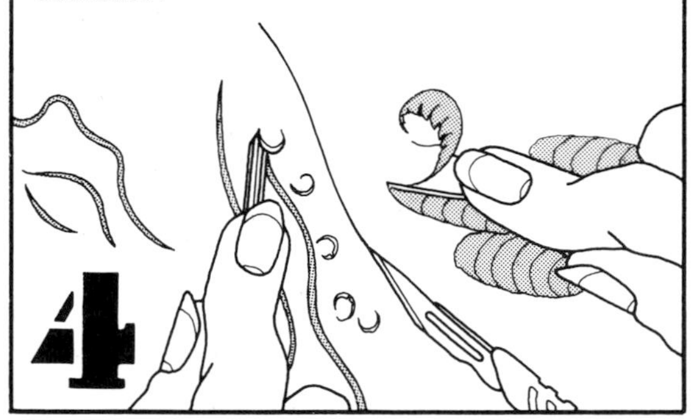

4

When taking a print, first roll out the required color of lino or block-printing ink on a sheet of glass. Then charge the linocut with the roller, making sure that all its raised surfaces are ink coated.

5

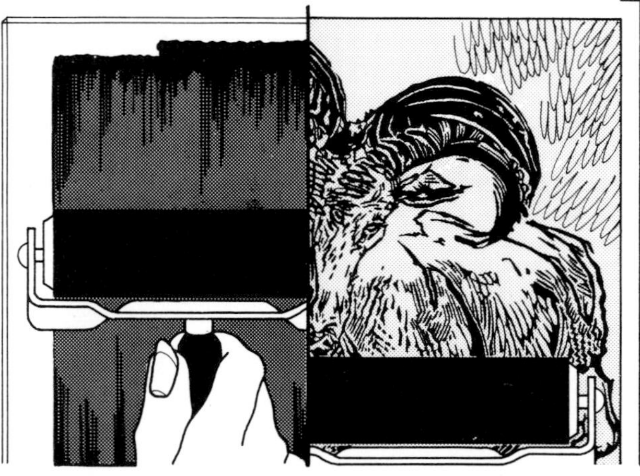

6

An impression is taken by laying a sheet of print paper over the inked-up face of the linocut and covering that with a protective sheet of paper, before submitting it to a small hand press or burnishing the back of the paper with the heel of the hand or a clean roller.

N.B.: When burnishing, make sure not to slide the print paper out of register with the block.

Registering Multicolored Linocuts

1

When a multicolored linocut is planned, it is necessary to produce one block for each color in the sequence. Accurate registration between the different colors during proofing requires that each lino block be trimmed exactly to the same size as the others, prior to receiving its incised artwork. The size of the block is determined by that of the master design.

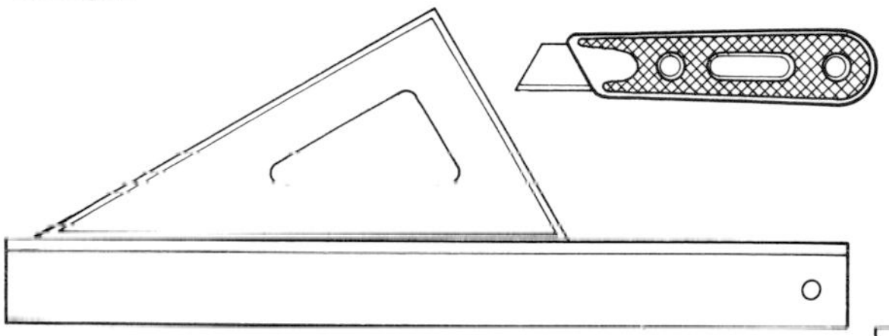

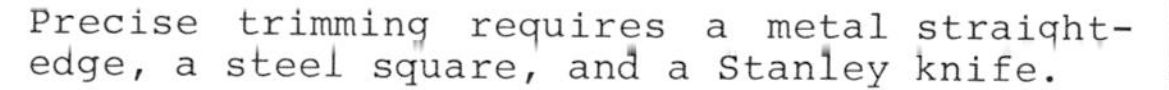

Precise trimming requires a metal straight-edge, a steel square, and a Stanley knife.

2

When trimming each block, begin by cutting one true edge. For registration purposes, this should be the bottom edge of the block. Then, after cutting the two sides accurately at right angles, make sure that the remaining top edge is cut parallel to the bottom edge.

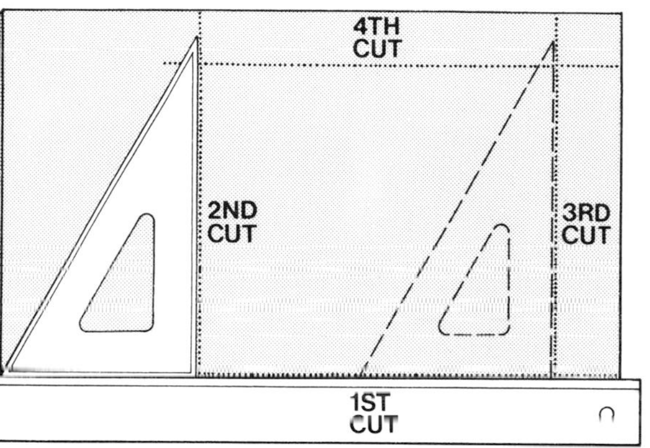

3

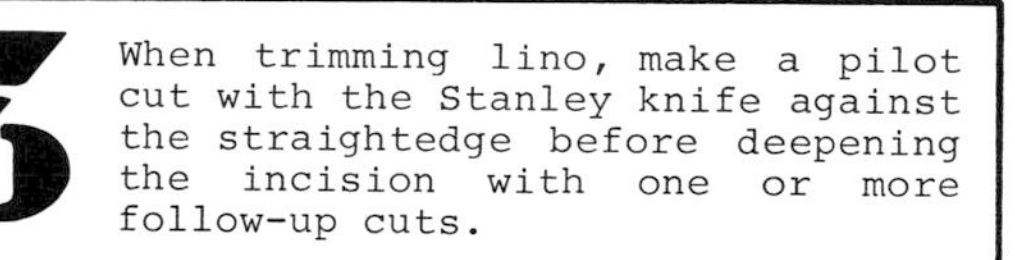

When trimming lino, make a pilot cut with the Stanley knife against the straightedge before deepening the incision with one or more follow-up cuts.

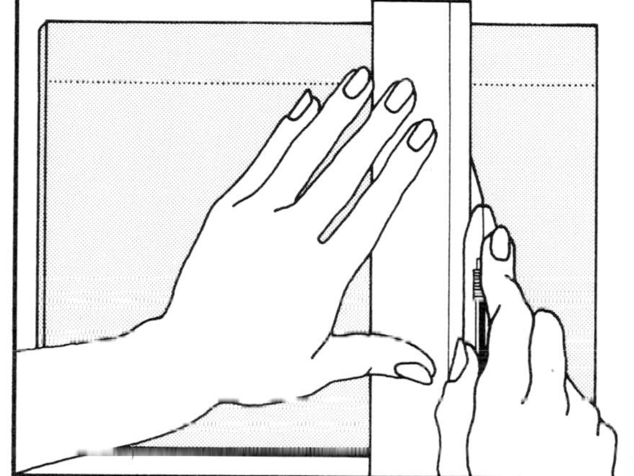

4

Thon fold baok thc lino away from the cut, so that any remaining thickness breaks along its length. Finally, working from the back, cut through the layer of backing canvas.

N.B.: Because the bottom and the left-hand edge of each block are critical to registration, make sure that these cuts are trimmed as cleanly as possible.

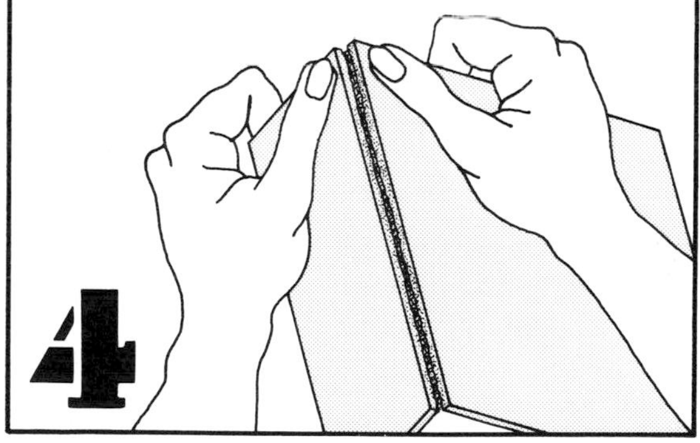

5

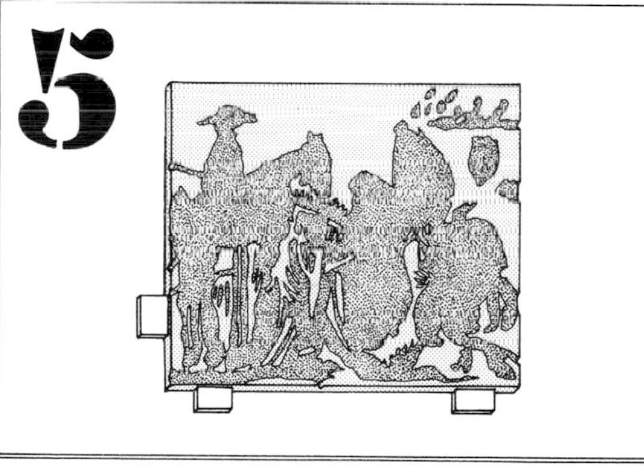

Registration begins after the master design has been traced onto the blocks and each has received its color-representative gouged artwork.

Place the first block at the center of a large sheet of cardboard. Then register its location by gluing three cardboard or Masonite tabs so that they abut the block--two along the bottom edge and one at the lower end of its left-hand side.

6

Finally, center a sheet of print paper over the block, then register its position by gluing three more tabs that key the paper to the block. In this fashion, each block--charged with its appropriately colored ink--can now be introduced accurately to thc print paper to take its impression.

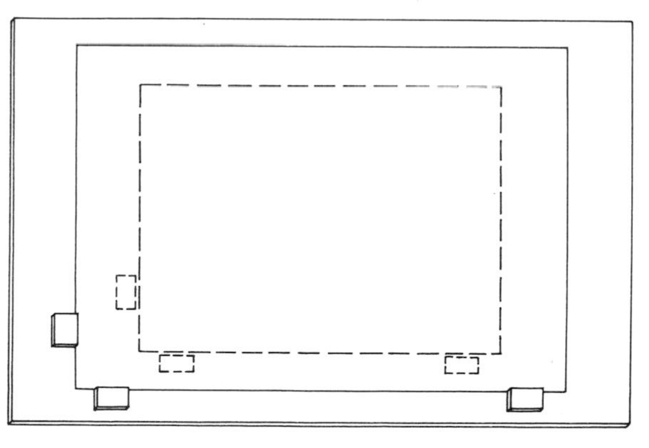

How to Make a Relief Etching

Although metal plates are normally employed for the various methods of intaglio printing, they can also be used for relief printing. For example, this relief etching technique is an excellent means of exploring a whole range of effects using the back of an old lithography plate.

First, brush apply a variety of shapes, using "stopping out" varnish, soap, wax, or a strongly adhesive grease.

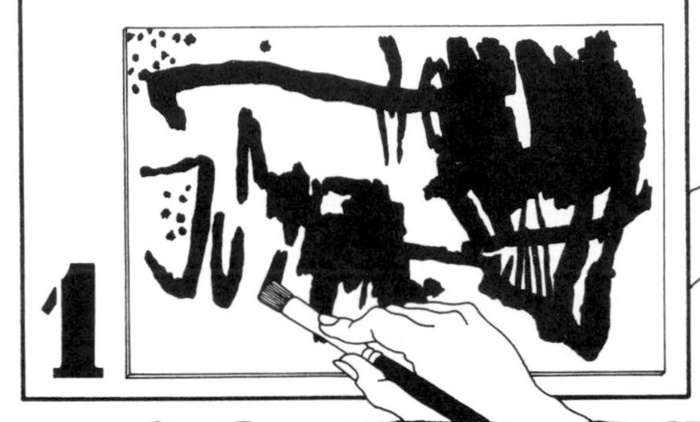

1

Next, draw into the shapes with a needle, matchstick, comb, or any other device that will cut back and expose the metal of the plate.

N.B.: Once the design is finished, the back and the side edges of the plate should be coated with a protective layer of "stopping out" varnish.

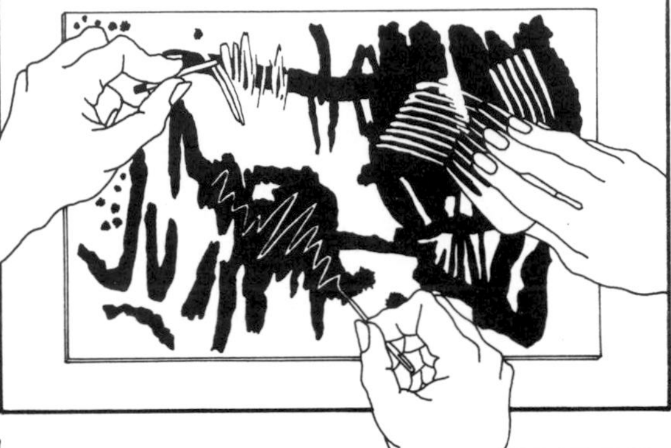

Then place the plate face up in a plastic or porcelain photographer's developing tray, and pour a solution composed of 1 part nitric acid to 12 parts of water onto its surface. If left submerged for about ten to fifteen minutes, the acid will eat sufficiently into the exposed areas of metal to produce in relief those areas protected by the acid-resistant varnishes.

N.B.: A plate requiring a deep bite will need a stronger etch solution, such as 1 part nitric acid to 7 parts of water. Apart from the strength of the acid and the duration of its exposure to the plate, the temperature of the acid also affects its capacity to etch--a warm solution being more effective than a cold one. However, the beginner is strongly advised to work initially under the direction of a tutor. Also, the making of etching solutions and the handling of plates should only be attempted after a thorough familiarization with studio processes.

4

An alternative method is to build a wall of Plasticine or putty around the edge of the plate so that the drawn surface itself becomes the tray into which the acid solution is poured.

N.B.: Extreme care should be taken when handling the acid. Keep your face well away from its fumes, and immediately wash away any splashes on skin or clothing with plenty of water. Always store the solution securely in a clearly marked bottle.

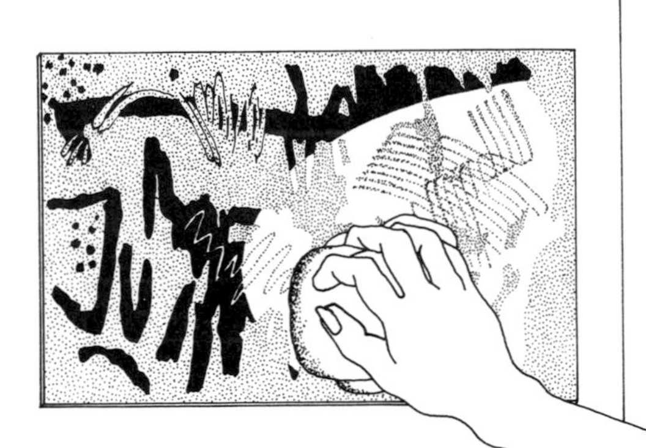

After removing the plate from the acid with a plate lifter, wash it under running water. Then, after drying the plate with blotting paper, place it face up on a clean sheet of paper and clean away the varnish, wax, and grease with a pad soaked in turpentine.

5

6

Make sure the plate is clean and dry before inking its surface, with litho ink and a roller.

Finally, the proofing stage can be achieved by placing a sheet of print paper over the inked surface and passing it through a mangle with rubber rollers. Otherwise the print can be taken by working over the back of the print paper on the plate with a clean roller.

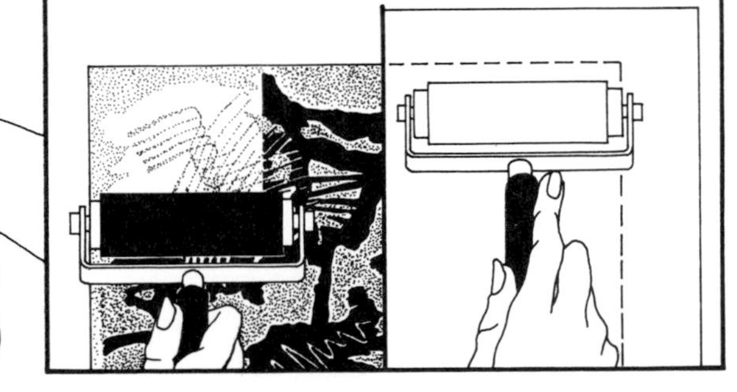

How to Etch a Design on a Sheet of Glass

Using a commercial etching fluid, it is possible to produce a design on a sheet of glass that can be used in exhibitions or in displays that are lit from behind. As there are many types of glass, it is wise to experiment with the etching fluid on a small piece before attempting a large design. However, as this process involves mild acid, great care should be exercised. Spillages on skin should be promptly washed away with plenty of water.

1

Prior to etching, the glass should be cleansed of grease and finger marks by washing it with a detergent, then drying it off with a clean, soft cloth. A stencil is then required to mask those areas of the glass that are not to be etched. Custom gum-backed stencil paper can be used, but a good mask is achieved by using soft-sheet contact plastic, such as that purchased in rolls for laminating table-tops and the like.

2

The design positive is then cut away from the plastic or paper stencil, using a scalpel. If centerpiece elements are involved, such as the middle of the letter O, these should be retained for repositioning them later on the glass.

3

Next, peel away the backing sheet from the stencil and locate it carefully on the glass. To ensure that all edges are firmly secured, smooth them with the back of a spoon or a small roller.

N.B.: If using more than one mask, be sure to align them. Also, as overlapping points are vulnerable to seepage, make sure they are well compressed.

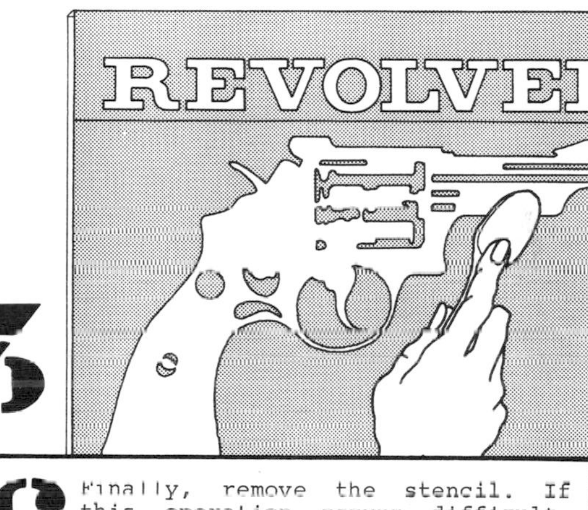

4

Then, after checking the instructions on the etching fluid's container, apply the fluid as thickly as possible to the exposed areas of the glass, using a cotton bud on the end of a toothpick.

N.B.: It is important that the glass be horizontal during this application.

5

Leave the fluid to activate on the glass for five to ten minutes, the exact "exposure" being dictated by the trial experiment. Then wash away the fluid with warm running water. Do this quickly, to avoid damage to the stencil. If, after drying the glass with a cloth, you see that areas require further etching, make sure the glass is completely dry before reapplying the fluid.

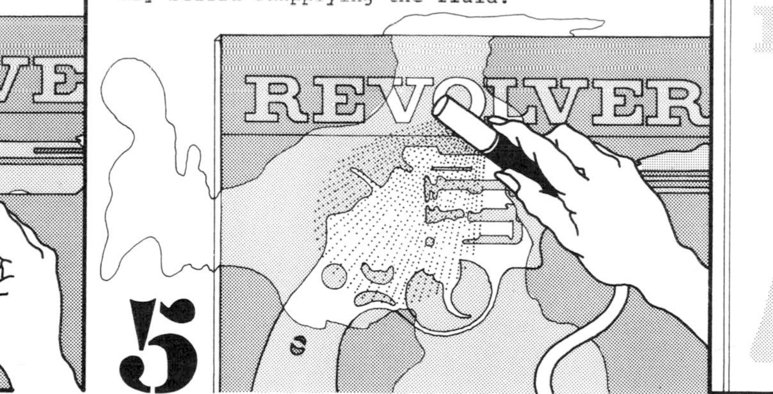

6

Finally, remove the stencil. If this operation proves difficult, soak the glass in warm water; if it still proves intractable, try using lighter fluid to loosen the glue.

Screenprinting: How to Stretch the Mesh

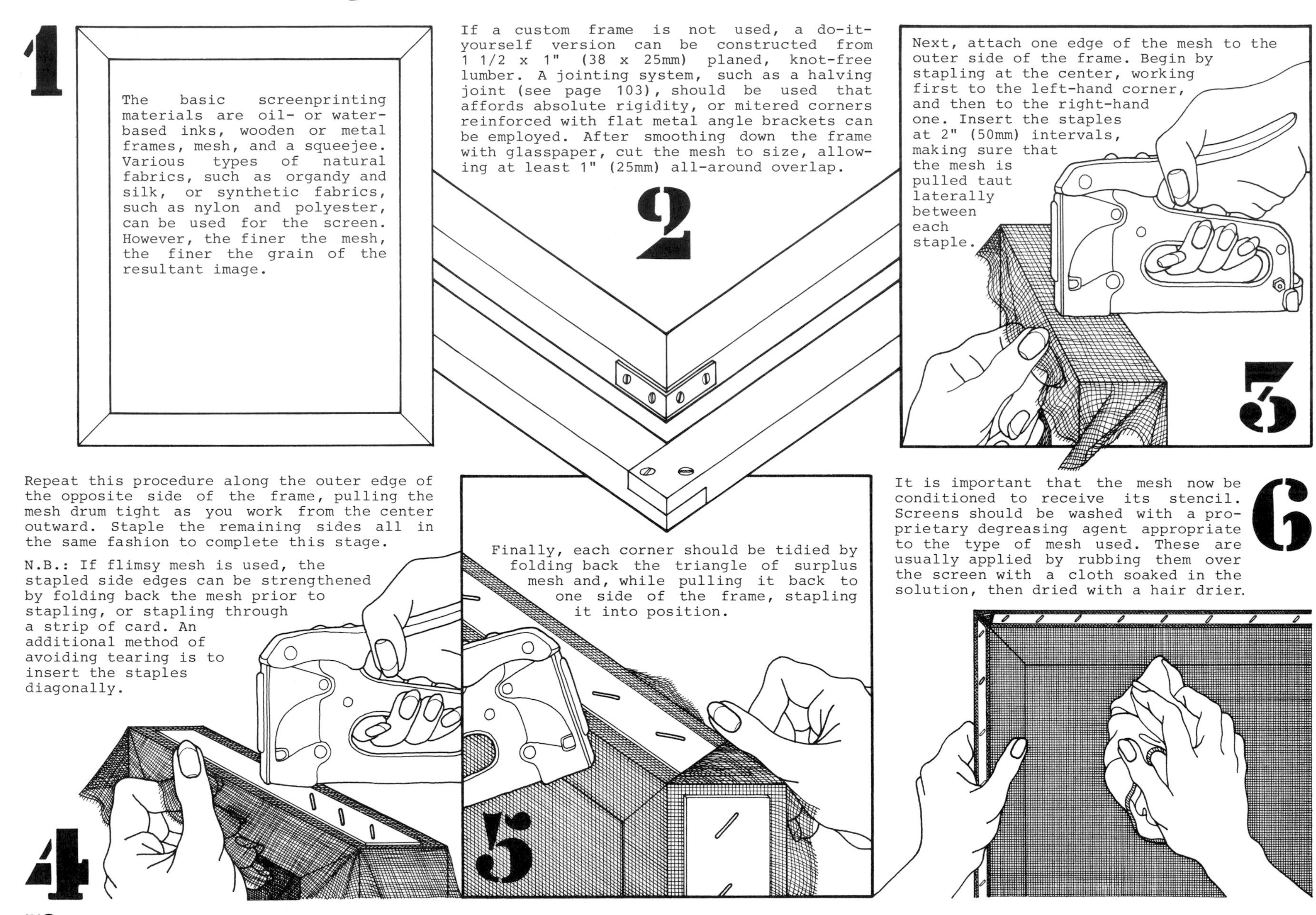

1 The basic screenprinting materials are oil- or water-based inks, wooden or metal frames, mesh, and a squeejee. Various types of natural fabrics, such as organdy and silk, or synthetic fabrics, such as nylon and polyester, can be used for the screen. However, the finer the mesh, the finer the grain of the resultant image.

2 If a custom frame is not used, a do-it-yourself version can be constructed from 1 1/2 x 1" (38 x 25mm) planed, knot-free lumber. A jointing system, such as a halving joint (see page 103), should be used that affords absolute rigidity, or mitered corners reinforced with flat metal angle brackets can be employed. After smoothing down the frame with glasspaper, cut the mesh to size, allowing at least 1" (25mm) all-around overlap.

3 Next, attach one edge of the mesh to the outer side of the frame. Begin by stapling at the center, working first to the left-hand corner, and then to the right-hand one. Insert the staples at 2" (50mm) intervals, making sure that the mesh is pulled taut laterally between each staple.

4 Repeat this procedure along the outer edge of the opposite side of the frame, pulling the mesh drum tight as you work from the center outward. Staple the remaining sides all in the same fashion to complete this stage.

N.B.: If flimsy mesh is used, the stapled side edges can be strengthened by folding back the mesh prior to stapling, or stapling through a strip of card. An additional method of avoiding tearing is to insert the staples diagonally.

5 Finally, each corner should be tidied by folding back the triangle of surplus mesh and, while pulling it back to one side of the frame, stapling it into position.

6 It is important that the mesh now be conditioned to receive its stencil. Screens should be washed with a proprietary degreasing agent appropriate to the type of mesh used. These are usually applied by rubbing them over the screen with a cloth soaked in the solution, then dried with a hair drier.

Preparing the Mesh

High quality screenprinted images, and their precise registration during overprinting, rely upon taut, well-stretched screens and their mesh being given an appropriate conditioner prior to use.

The best method of pretreating a newly stretched mesh is to use a proprietary universal degreasing paste. This conditions the mesh by removing any oils from its fiber and, as an aid to the adhesion of stencil film, raises a slight tooth on the surface of the screen. This treatment is also recommended for used screens, as it removes any ink residue and other impurities left behind after screen reclamation (see page 87).

N.B.: Some marketed conditioners that abrade the mesh are cloth applied and then heat dried. A do-it-yourself method is to gently rub the mesh with a fine grade wet-and-dry abrasive paper. However, when using a custom product, always follow the manufacturer's instructions.

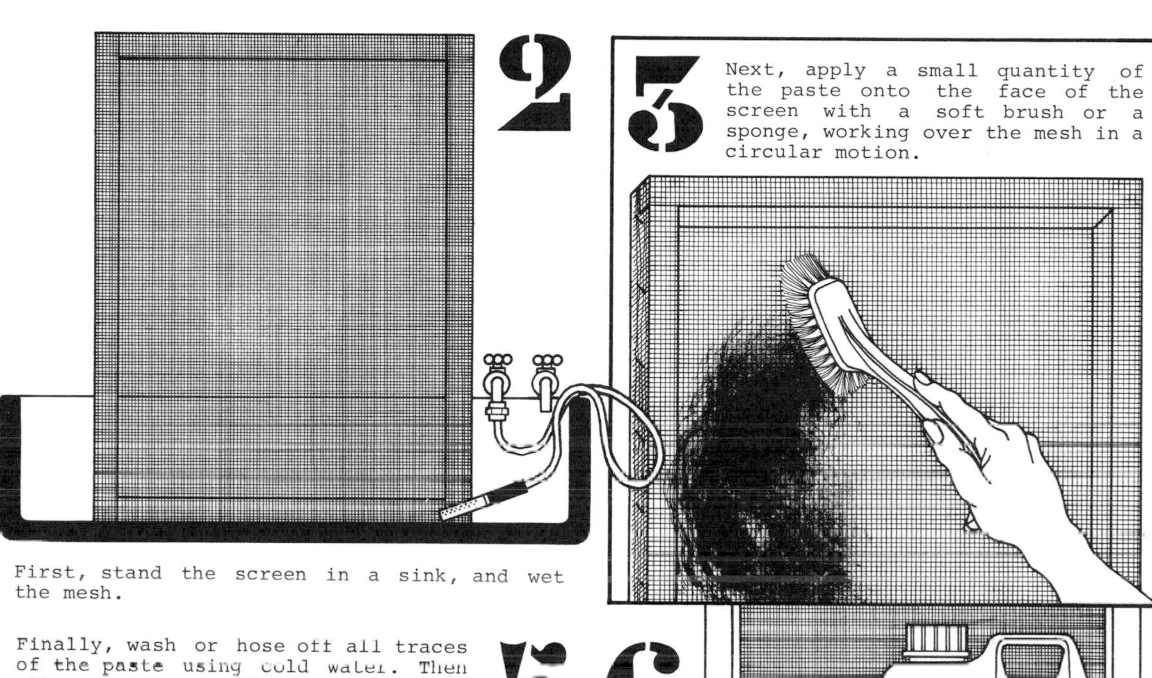

1 First, stand the screen in a sink, and wet the mesh.

2

3 Next, apply a small quantity of the paste onto the face of the screen with a soft brush or a sponge, working over the mesh in a circular motion.

4 Repeat the same operation on the reverse side of the mesh. When finished, leave the screen for a few minutes in order to allow the solution to react.

5 Finally, wash or hose off all traces of the paste using cold water. Then allow the screen to dry thoroughly prior to tape or stencil application.

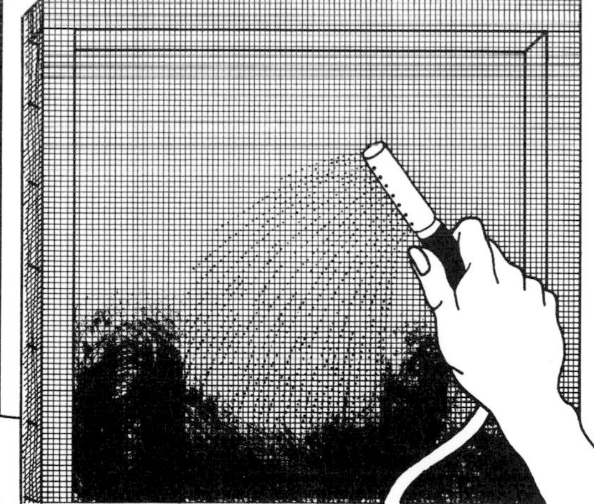

6

There is also a proprietary concentrated liquid detergent that is used to degrease a mesh not requiring an abraded surface, such as that receiving direct emulsions (see pages 80-81). This treatment reduces pinholes caused by dust particles by eliminating static, and improves the flow of emulsion during its application.

The application of the liquid detergent is identical to that of the paste.

N.B.: Although some degreasing agents are nonhazardous, others are caustic. Therefore, reasonable precautions should be taken during their use.

Taping the Screen and Making a Printing Bed

1

Before the screen is used, it should be taped to prevent ink loss between the frame and the mesh, and also to provide an ink "trough."

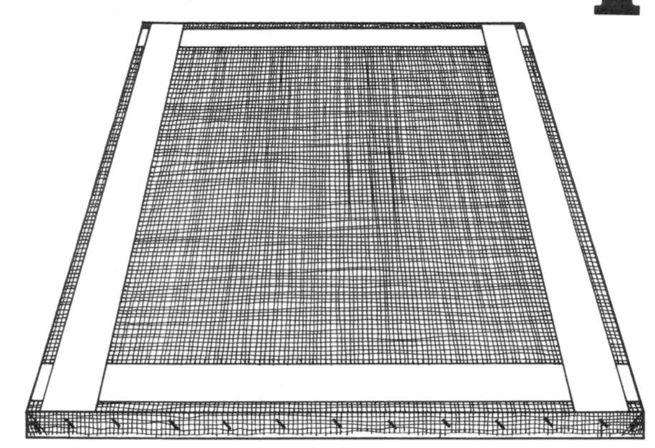

Using 2" (50mm) vinyl tape or gummed paper strip, tape the outer face of the screen, allowing half to overlap the frame, half to overlap the mesh.

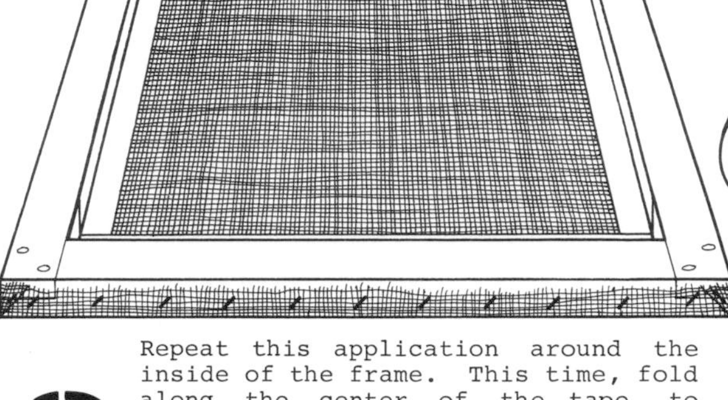

2

Repeat this application around the inside of the frame. This time, fold along the center of the tape, to assist in positioning half the tape against the inside edge of the frame and half against the inside edge of the mesh.

3

Add two extra pieces of tape to extend those at each of the narrow ends of the mesh. These will act as ink "troughs." The screen is now ready to receive its stencil after it is pretreated with a degreasing agent appropriate to the stencil used. In order to ascertain the correct preparation, consult the product information accompanying each proprietary stencil type.

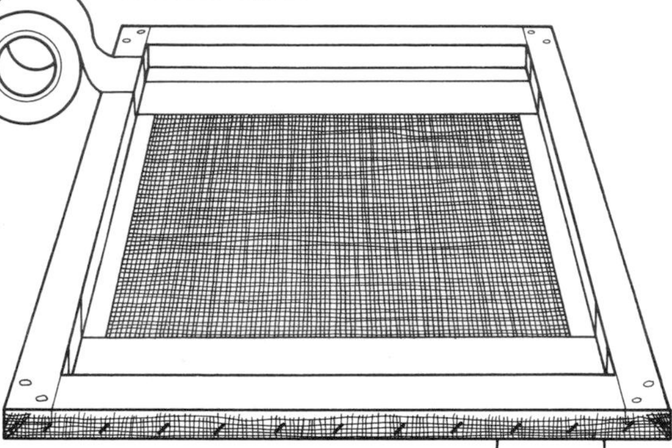

4

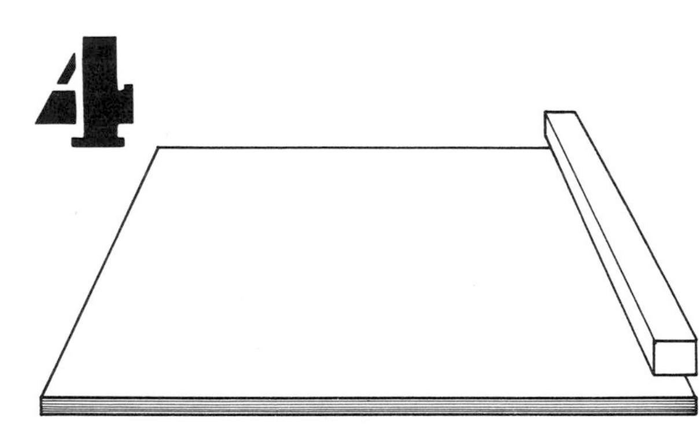

If multicolored print runs--requiring a separate screen for each color--are planned, then a printing bed should be constructed prior to the screen-stretching stage. A printing bed consists of a blockboard or plywood baseboard cut with a 5" (125mm) margin greater than the size of the screen. A piece of wood with dimensions similar to that used for the screen frame is also required as a hinge bar, its length corresponding to the width of the baseboard.

5

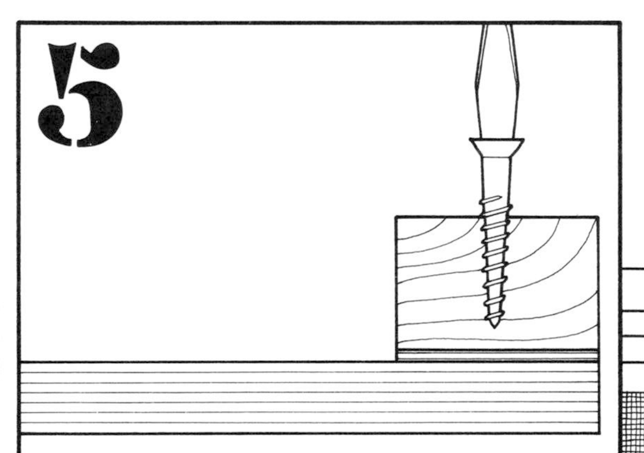

Screw the hinge bar into position along one end of the baseboard, with a packing of thick card sandwiched between the two.

The packing keeps the inked mesh off the paper during the printing operation, contact being made only by the tip of the moving squeegee.

6

Attach the screenprinting frame to the hinge bar using the type of hinge that allows its pin to be removed. This easy withdrawal and replacement of the hinge pin allows for rapid and accurate interchange of screens during printing.

How to Make Instant Multicolored Screenprints

1 A fast and exciting method of producing rapid screenprints is to work a design, such as that for a poster, directly onto a clean screen by drawing with oil pastels using as many colors as required.

N.B.: A full range of marks may be exploited, from fine lines to solid blocks of color.

2 Next, mix a solution comprising half turpentine and half silkscreen extender medium.

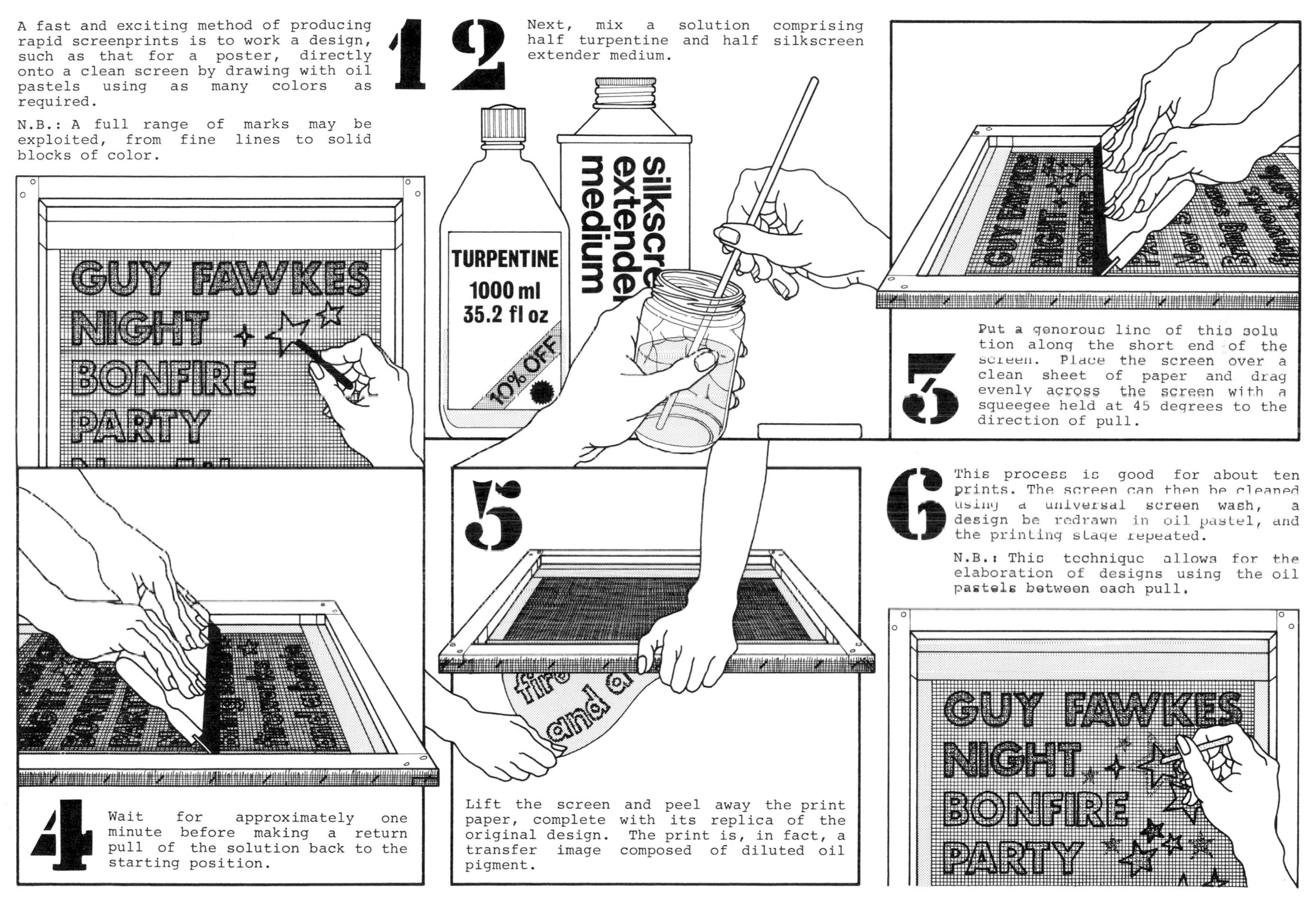

3 Put a generous line of this solution along the short end of the screen. Place the screen over a clean sheet of paper and drag evenly across the screen with a squeegee held at 45 degrees to the direction of pull.

4 Wait for approximately one minute before making a return pull of the solution back to the starting position.

5 Lift the screen and peel away the print paper, complete with its replica of the original design. The print is, in fact, a transfer image composed of diluted oil pigment.

6 This process is good for about ten prints. The screen can then be cleaned using a universal screen wash, a design be redrawn in oil pastel, and the printing stage repeated.

N.B.: This technique allows for the elaboration of designs using the oil pastels between each pull.

Making Two-color Paper Stencil Screenprints

Screenprinting involves the pressing of ink through designated areas of a stretched mesh so that a required design is printed onto paper placed immediately below the screen. The "negative" areas of a design, that is, those areas of the mesh not intended to be printed, must therefore be masked with a stencil.

A basic stencil is made by cutting predetermined shapes from a sheet of thin paper that expose the "positive" areas of the mesh to the printing ink.

N.B.: Newsprint can be used with spirit-based inks, but wax Kraft paper is recommended for better results, particularly in conjunction with water-based inks.

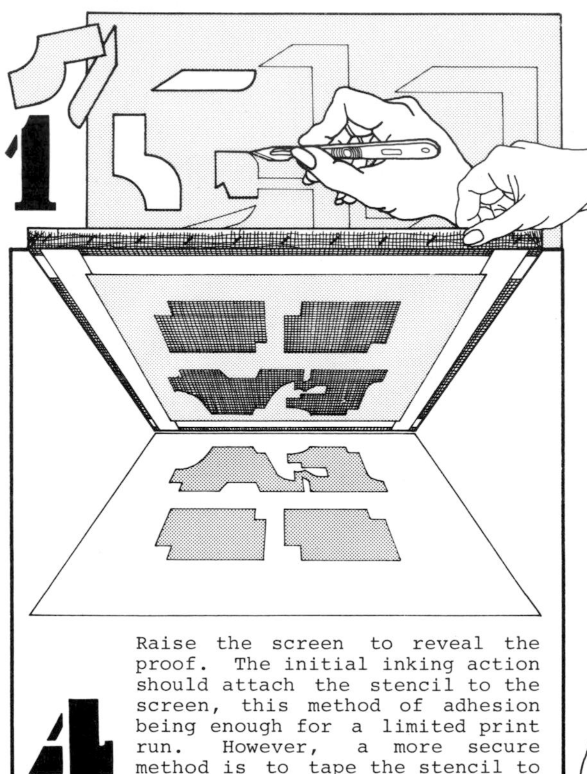

1

4 Raise the screen to reveal the proof. The initial inking action should attach the stencil to the screen, this method of adhesion being enough for a limited print run. However, a more secure method is to tape the stencil to the screen.

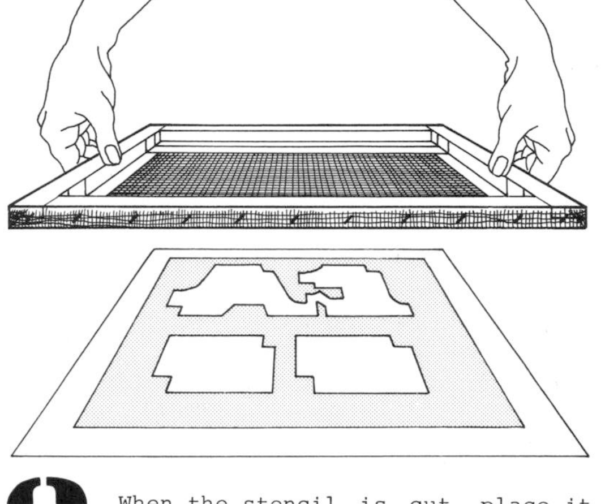

2 When the stencil is cut, place it on top of a sheet of printing paper, and then place the screen face down over the stencil.

3 Next, introduce the ink to the "trough," and pull the squeegee firmly across the mesh.

5 For a two-color print, a second paper stencil is prepared. This is placed on top of, and registered with, the first color print and, with a second color of ink, the printing process repeated.

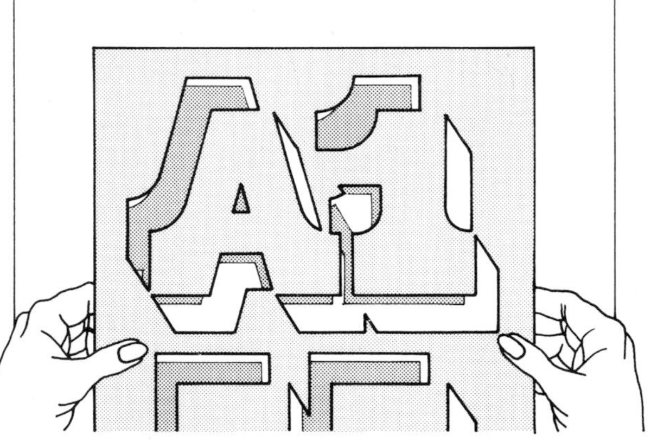

6 A third color--achieved by overlaying one ink on another--will not occur if opaque ink is used for overprinting. To obtain transparency, add an extender medium to the second ink mix, or use the "process" colors.

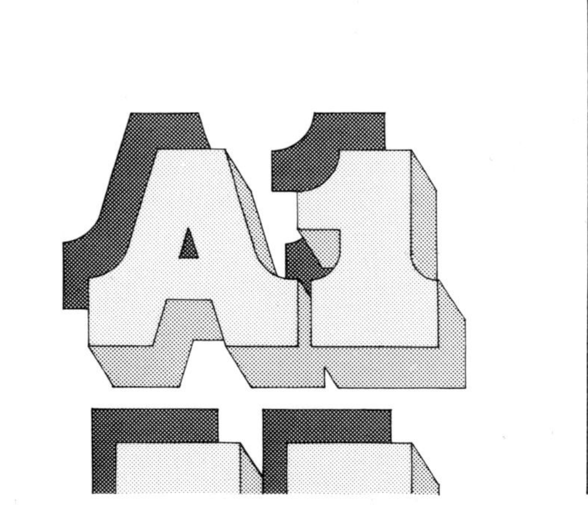

Other Methods of Making Screenprint Stencils

1

Another basic method of blocking the mesh is to draw a positive design directly and firmly onto the screen with oil pastel. When printed, this technique produces richly textured negatives, overprints being obtained by a further blocking of the mesh between color prints.

N.B.: As oil pastel is soluble in turpentine, a water-based ink must be used.

2

Oil pastel can also be used to achieve positive stencils, by being applied directly to the areas of the mesh that will ultimately print.

3

After application of the oil pastel, one entire side of the mesh is coated with a proprietary liquid screen filler medium and, when dry, the oil pastel is washed out from the other side with turpentine to reveal the positive areas of open mesh.

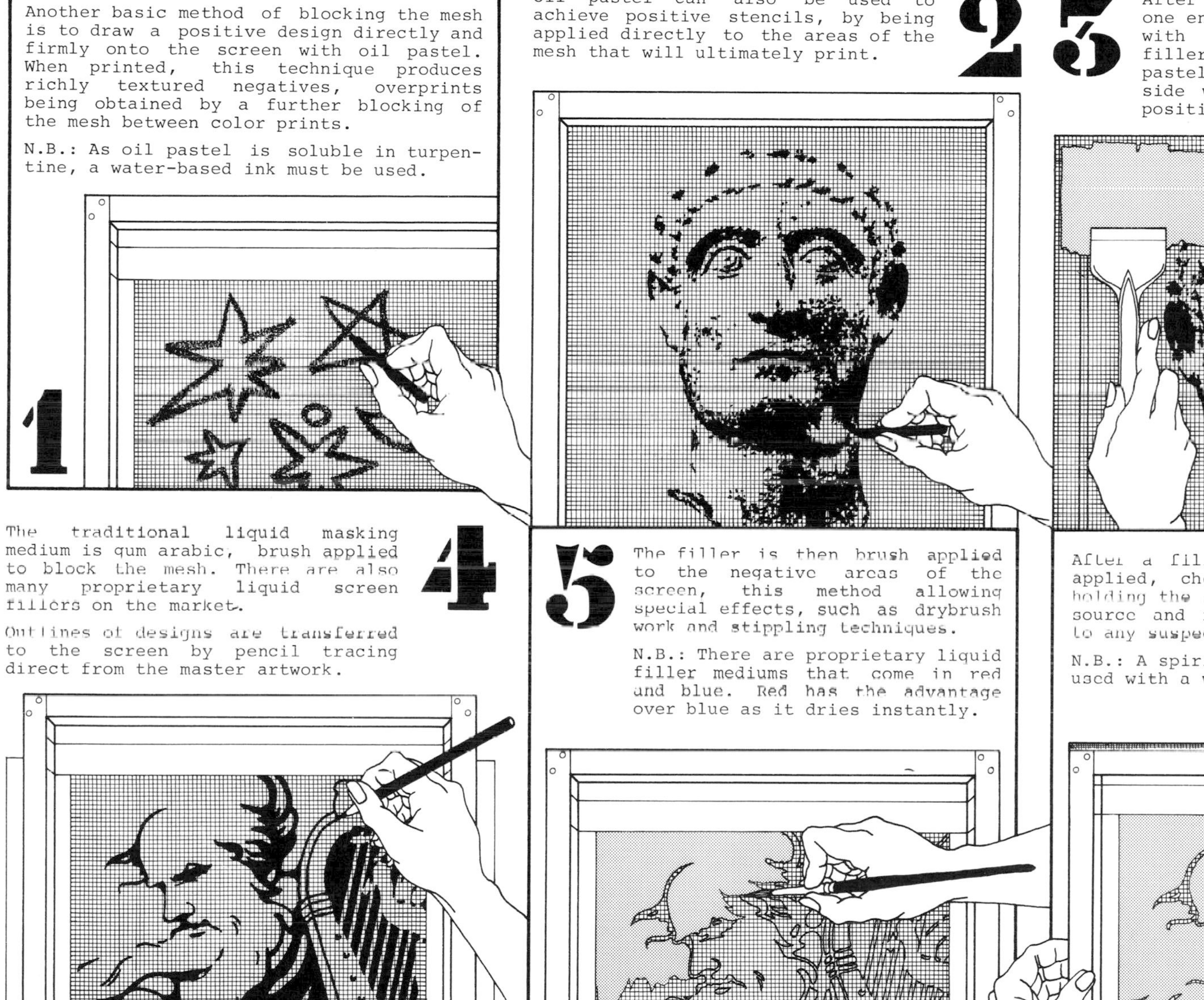

4

The traditional liquid masking medium is gum arabic, brush applied to block the mesh. There are also many proprietary liquid screen fillers on the market.

Outlines of designs are transferred to the screen by pencil tracing direct from the master artwork.

5

The filler is then brush applied to the negative areas of the screen, this method allowing special effects, such as drybrush work and stippling techniques.

N.B.: There are proprietary liquid filler mediums that come in red and blue. Red has the advantage over blue as it dries instantly.

6

After a filler medium has been applied, check for pinholes, by holding the screen against a light source and reapplying the medium to any suspect areas.

N.B.: A spirit-based ink should be used with a water-soluble stencil.

The Autocut Stencil Method

The autocut stencil combines a film of gelatin superimposed on a layer of polyester, which acts as a backing sheet.

1 First, place the stencil over the original design to be printed, with the side carrying the backing sheet face down over the artwork.

2 Then, using a sharp scalpel, trace cut the areas to be printed--making sure that only the top layer of gelatin is penetrated by the blade.

N.B.: A good tip is to work the cutting operation on a sheet of glass, as this avoids damage to the backing sheet.

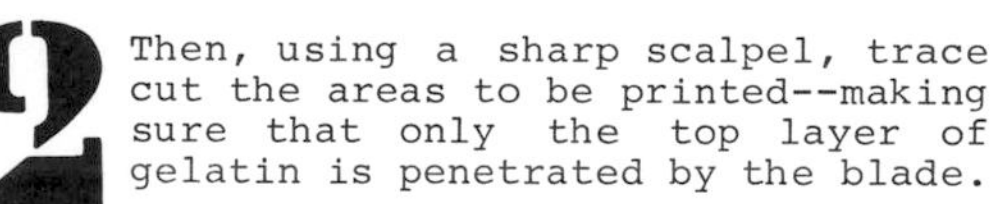

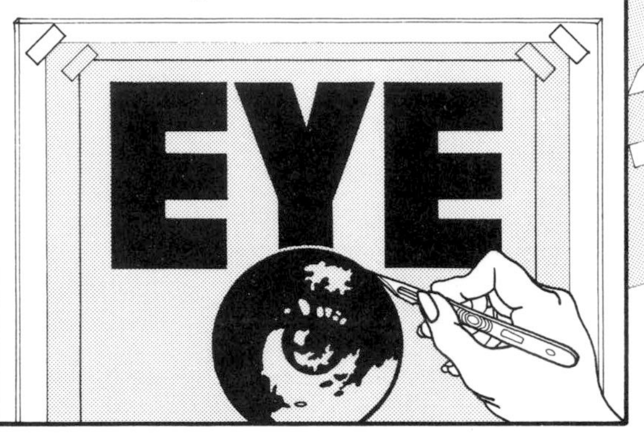

3 When the cutting stage is completed, remove the unwanted shapes of gelatin from the stencil after carefully lifting their edges with the scalpel blade.

4 Next, place the prepared stencil face up on a sheet of clean card with dimensions slightly less than those of the inside of the screen. Lay the screen face down over the stencil, making sure that the stencil is centered on the mesh and that there is good contact between them.

5 Wet a sponge in clean water and wipe evenly across the screen until the stencil adheres to the mesh. Make sure that air pockets are eliminated before quickly blotting off any excess water with one or two applications of absorbent paper.

N.B.: Be careful not to overblot, as this can cause the edges of the stencil to spread.

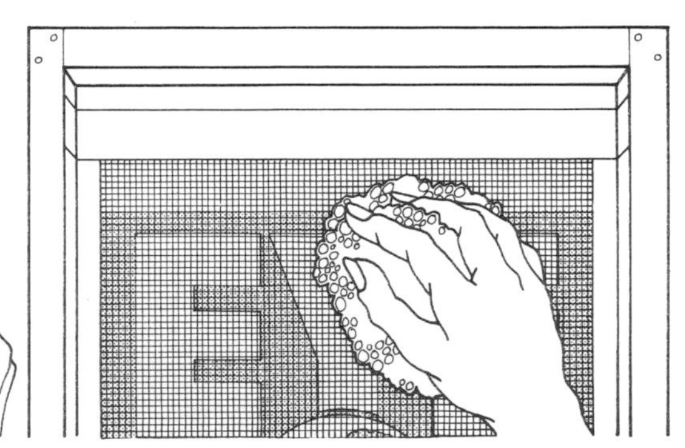

6 There are several types of knife-cut stencil film. Some are adhered with a mixture of approximately three parts water to one part methylated spirits, a mixture that reduces the drying time required.

Other stencils are transferred with a hot iron, so make sure that the correct adhesion method is used.

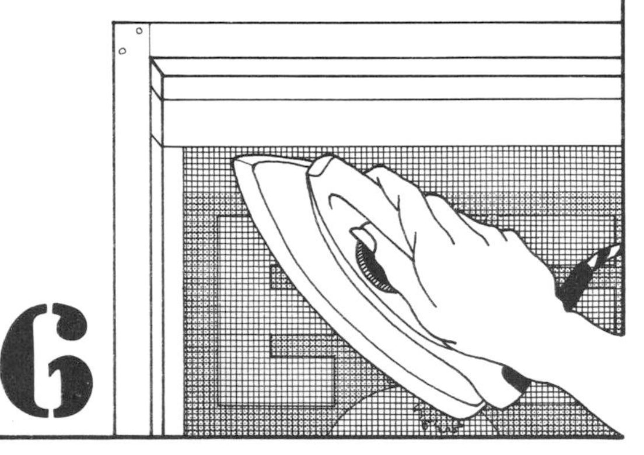

The Autocut Stencil Method

7 If water has been used in adhering the stencil to the mesh, dry the screen by supporting it in an upside-down position on two stools and above a portable fan heater emitting warm air for about twelve minutes.

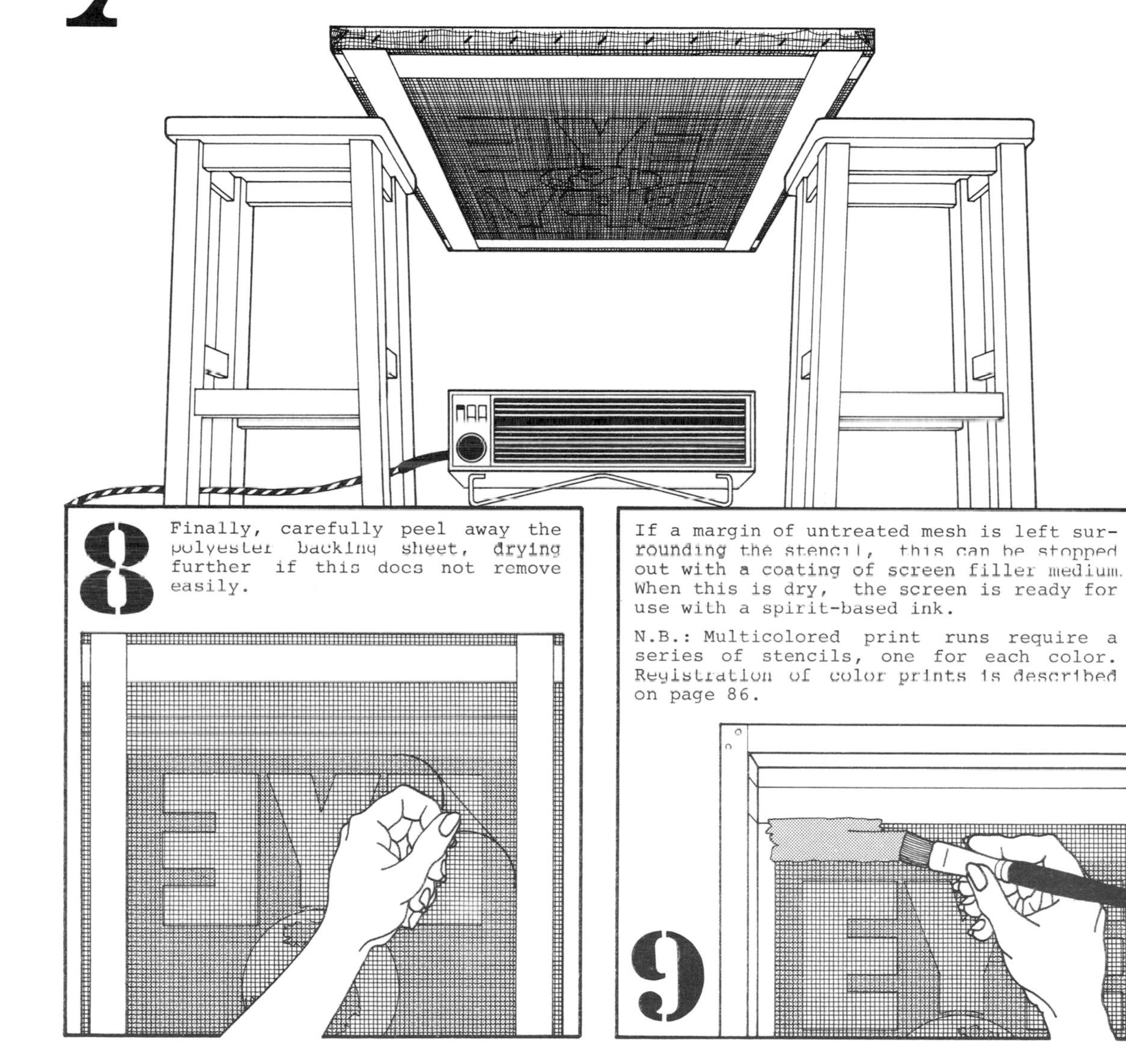

8 Finally, carefully peel away the polyester backing sheet, drying further if this does not remove easily.

9 If a margin of untreated mesh is left surrounding the stencil, this can be stopped out with a coating of screen filler medium. When this is dry, the screen is ready for use with a spirit-based ink.

N.B.: Multicolored print runs require a series of stencils, one for each color. Registration of color prints is described on page 86.

10

These are details from two posters screenprinted using knife-cut stencils--the upper example using one color and the lower example using two colors--by Waterman Graphics of Charlbury, Oxfordshire. Although highly intricate designs can be achieved using knife-cut stencils, the amount of time required for their preparation restricts their commercial use to designs comprising basic letter forms and shapes.

The Photostencil Screenprint (Direct Method)

1

The direct photostencil method is ideal for those tackling photographic screenprinting for the first time. It relies upon an opaque positive image being superimposed over a screen coated with a proprietary twin- or triple-pack light-sensitive medium comprising a mixture of emulsion, sensitizer, and colorant.

The stencil can be either a photographic film--positive printed onto film instead of paper--or a drawing worked in technical pen, grease pencil, or black acrylic paint, applied to clear plastic film or tracing paper. Alternatively, the stencil can be an actual flat object.

N.B.: When using the photostencil method, the mesh needs to be degreased. This is done prior to taping, using a proprietary degreasing agent or methylated spirits, then being allowed to dry (see page 73).

2

Working away from any direct light source, apply an even coating of the premixed emulsion using a coating trough. The coating trough is the recommended applicator as it transfers an even layer of the medium to the screen.

N.B.: It is crucial that all handling of the emulsion prior to its exposure be carried out under illumination low in blue or ultraviolet content, such as yellow fluorescent or weak tungsten.

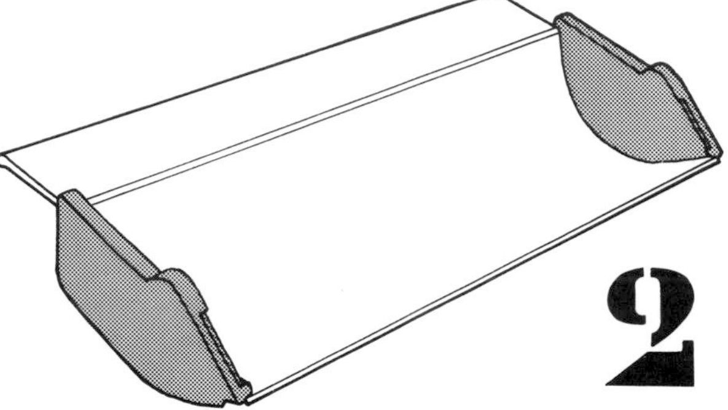

3

When applying the emulsion, stand the screen on edge and tilt slightly while drawing the loaded coating trough upward across the mesh. Apply two or three coats--the number depending upon the product requirements of the emulsion--to both sides of the screen.

4

Alternatively, the emulsion can be applied using a thick piece of card. A thick coat is applied to both sides of the screen, later scraping each side to remove any surplus.

N.B.: To avoid problems during developing, it is important to achieve an even coating of the emulsion on the mesh.

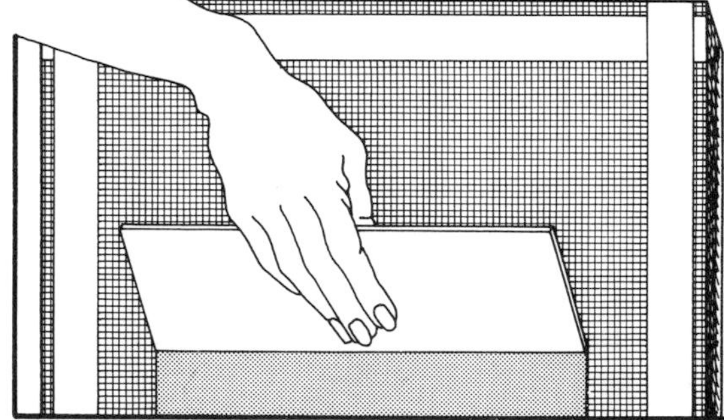

5

Next, dry the coated screen in a horizontal position in a darkened room with the gentle heat of a warm air fan, taking care not to blow dust onto the emulsion.

6

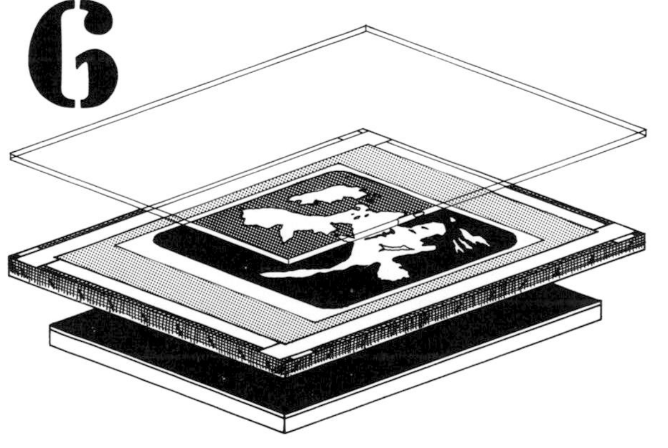

Before exposure, and with the screen positioned face up, place the film, positive emulsion side down, in intimate contact with the emulsion on the coated screen. Then place the screen face up on a raised pad topped with black paper. To ensure that the film positive is in direct contact with the coated surface of the mesh, cover the assembly with a sheet of clean glass.

The Photostencil Screenprint (Direct Method)

7

The next stage involves exposing the screen to light. Light sources for exposure range from mercury vapor to direct sunlight. Artificial illuminants should be positioned two feet (600mm) above the screen. Exposure times vary and depend on the nature of the light source, the thickness of the emulsion, the grade of mesh, and the quality of the stencil. The following exposure guide is based on medium, white mesh covered each side with three coats of Dirasol direct photo-stencil emulsion.

EXPOSURE GUIDE

Light source	Distance	Time
50 amp Open Carbon Arc	48" (1200mm)	16-18 mins
HPR 125 Mercury Vapor Lamp	20" (500mm)	8-9 mins
2000w Metal Halide	48" (1200mm)	6-7 mins
5000w Metal Halide	48" (1200mm)	1 3/4-2 mins

When using a do-it-yourself exposing setup, a trial exposure check on a small screen is recommended. This is done by placing a strip of clear film over the photostencil emulsion, allowing a series of stepped exposures using a black card mask--each exposure time doubling that of the last.

The correct exposure is achieved by that exposed step showing no difference in color density when compared with the surrounding, directly exposed area. For example, if the area exposed through the film is lighter than the surrounding area, then the screen has been overexposed.

8

After exposure, the screen is developed by standing it in a sink and gently spraying both sides with lukewarm water from a shower attachment or a hose connected to a faucet. After two minutes, increase the spray pressure slightly and continue washing out the soluble, unexposed areas of the stencil.

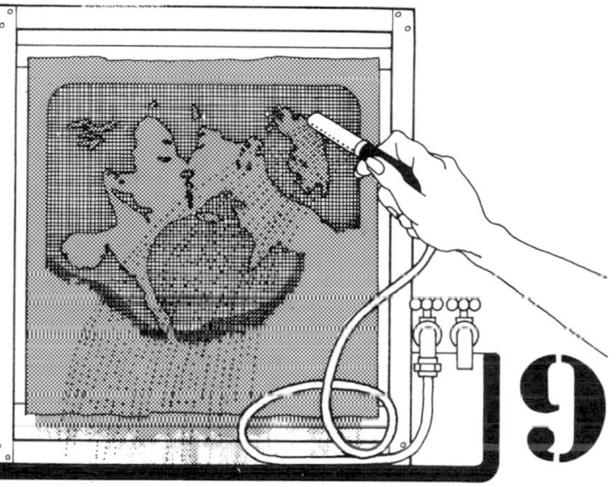

9

10

When the developed stencil image appears clean and sharp, remove the screen to a clean horizontal surface, and gently mop up any surface moisture from both sides of the mesh with a soft wash leather.

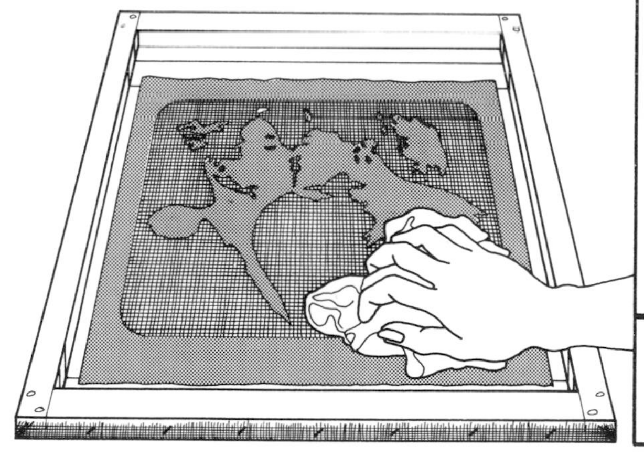

11

Complete the drying process by supporting the screen horizontally between two stools, and over the warm air of a portable fan heater.

Finally, check the screen against a light source for pinholes. If necessary, these can be filled in with a brush loaded with the emulsion, then reexposed to direct light.

N.B.: Pinholes can result from dirt specks on the glass or film positive during exposure, from a dirty coating trough, or from dust in the emulsion. It is important, therefore, to keep all equipment and material scrupulously clean.

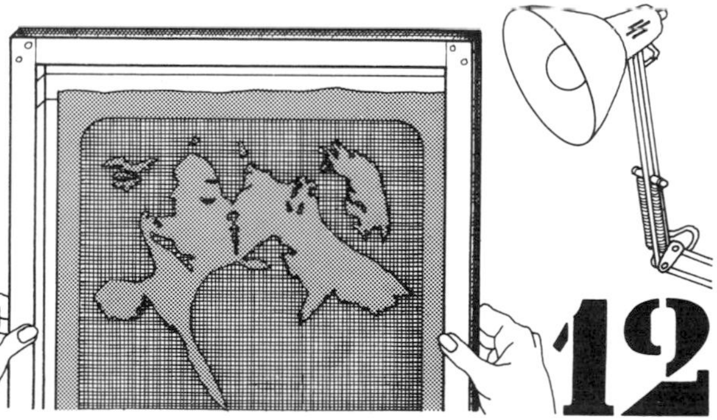

12

The Photostencil Screenprint (Indirect Method)

The indirect photostencil method, so called because the stencil is processed away from the screen, is very popular because of its ability to reproduce screenprinted versions of highly detailed halftone photographic images. There are many proprietary photostencil films on the market that are, basically, a layer of sensitized gelatin laminated to a transparent film plastic backing sheet.

N.B.: Correct storage of photostencil material is important. Make sure to keep it away from photographic papers and films, and from materials that might contain synthetic glues or preservative chemicals.

1 2

The process begins with a film positive. Because their quality dictates that of the ultimate print, positives should be in good condition, with any surface dust carefully wiped away with a clean, dry wash leather.

The stencil is also required at this stage. It should be removed from its container and handled only under subdued daylight or low-wattage tungsten or yellow fluorescent tubes. Unroll the stencil film on a flat surface and cut it to the required size, using a sharp blade to make sure not to wrinkle the film.

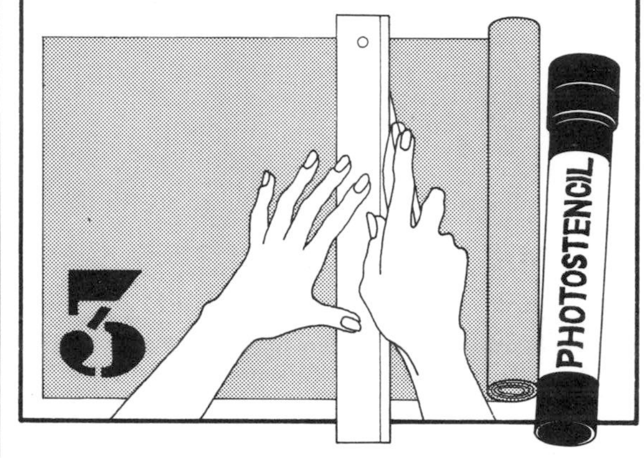

4

The film positive image is now transferred to the stencil material via an exposure unit. A do-it-yourself exposure unit can be assembled from a sandwich comprising a sheet of glass or acrylic placed over a sheet of black foam rubber supported on a drawing-board base. A mercury vapor lamp or a light source with ultraviolet content is positioned approximately two feet (600mm) above the glass sheet.

Place the film positive with its emulsion side upward onto the glass sheet of the exposure unit. Then overlay this with the photostencil, making sure that it is centered and that its plastic backing is face down and is in contact with the film positive.

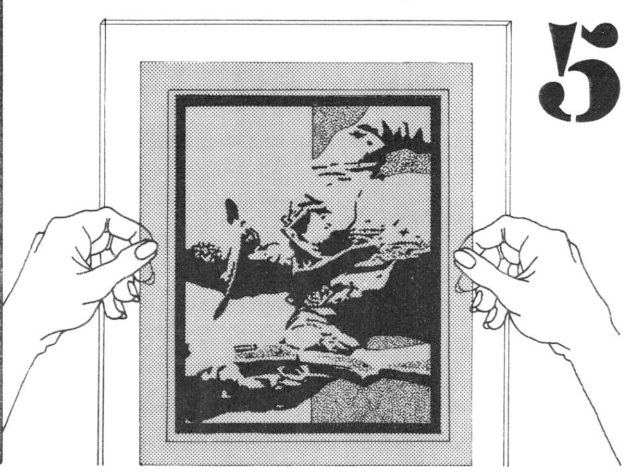

5

The Photostencil Screenprint (Indirect Method)

6 Before exposure invert the glass sheet, complete with the positive and the stencil, so that the latter sits underneath the film positive when the glass is replaced on the foam rubber bed.

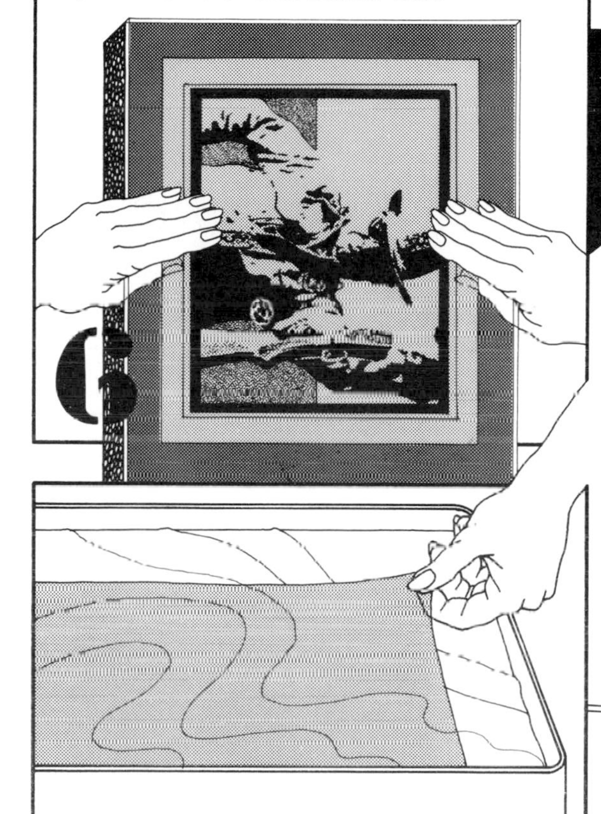

7 Switch on the lamp to begin the exposure. Exposure times vary, depending on both the types of stencils and the illumination. For example, this exposure guide is recommended for the Five-Star photostencil product:

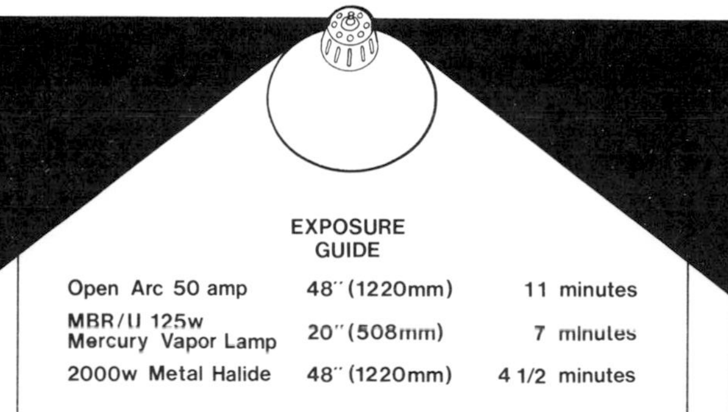

EXPOSURE GUIDE

Open Arc 50 amp	48" (1220mm)	11 minutes
MBR/U 125w Mercury Vapor Lamp	20" (508mm)	7 minutes
2000w Metal Halide	48" (1220mm)	4 1/2 minutes

N.B.: Overexposure can adversely affect the subsequent adhesion of the stencil to the mesh.

8 Immediately remove the stencil from the exposure unit and immerse it for one minute in a bath of proprietary developer at a temperature of 68°F (20°C). This immersion functions to harden the exposed, or negative, areas of gelatin.

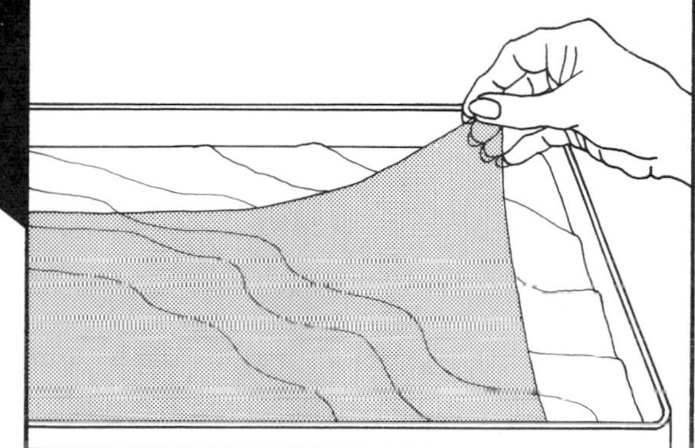

9 An alternative do-it-yourself developer can be prepared from a mixture of one part hydrogen peroxide (20-volume strength) and four parts water, the former being purchased from any drugstore. Immersion time in this solution and its temperature are the same as those for the custom developer.

10 Remove the stencil and hang it over a sink. The soft unexposed areas of the gelatin should now be washed out with lukewarm water sprayed from a shower attachment until the transferred image becomes clearly visible.

N.B.: An alternative method is to splash the stencil in a bath of lukewarm water. However, avoid overwashing, as screen adhesion may be impaired.

11 The stencil is now placed gelatin side down onto the underside face of the mesh.

N.B.: It is important that the screen have been pretreated with a degreasing agent.

The Photostencil Screenprint (Indirect Method)

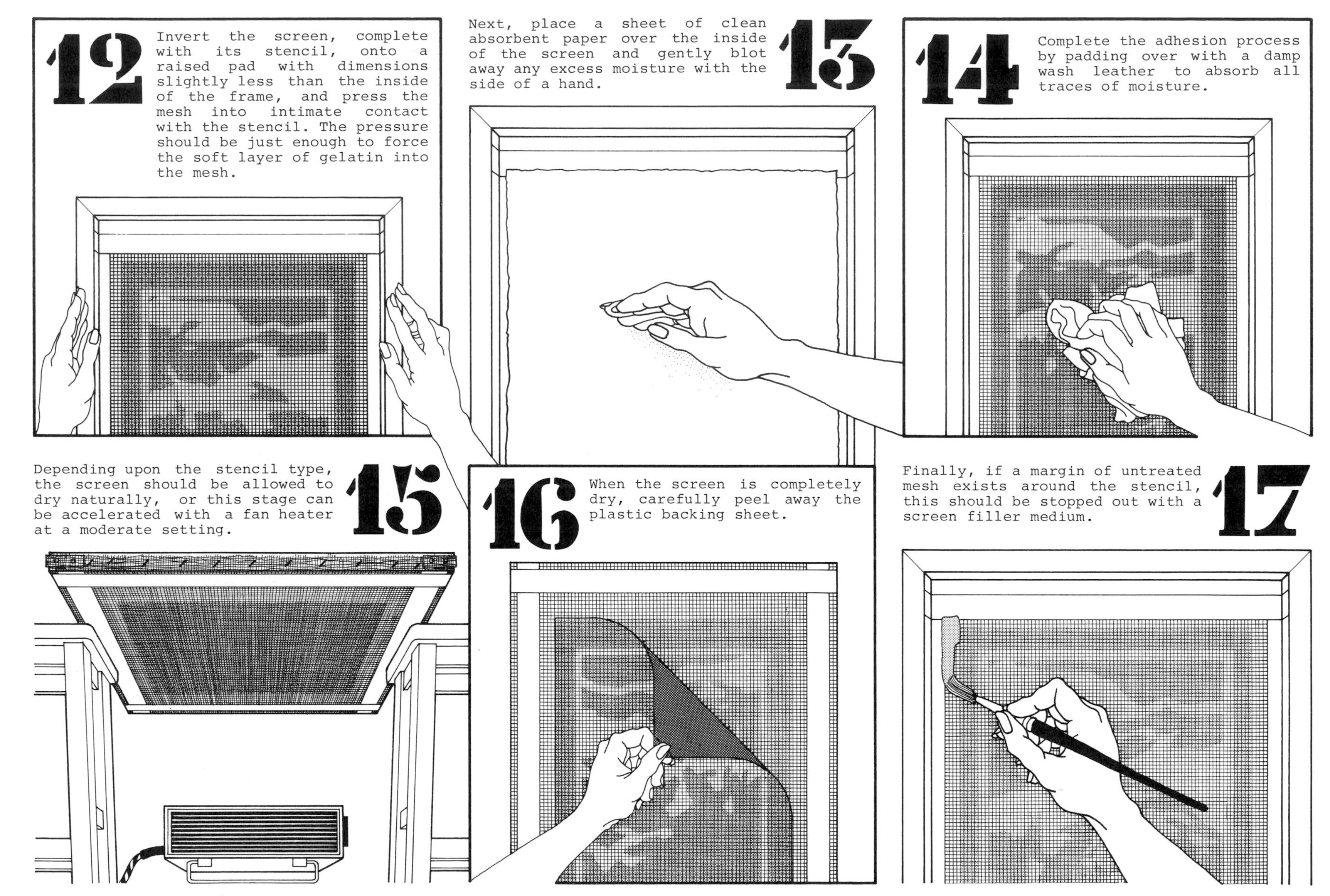

12 Invert the screen, complete with its stencil, onto a raised pad with dimensions slightly less than the inside of the frame, and press the mesh into intimate contact with the stencil. The pressure should be just enough to force the soft layer of gelatin into the mesh.

13 Next, place a sheet of clean absorbent paper over the inside of the screen and gently blot away any excess moisture with the side of a hand.

14 Complete the adhesion process by padding over with a damp wash leather to absorb all traces of moisture.

15 Depending upon the stencil type, the screen should be allowed to dry naturally, or this stage can be accelerated with a fan heater at a moderate setting.

16 When the screen is completely dry, carefully peel away the plastic backing sheet.

17 Finally, if a margin of untreated mesh exists around the stencil, this should be stopped out with a screen filler medium.

Indirect Photostencil Screenprint

This is an indirect photostencil screenprint using two screens, one with a film negative stencil and one with a film positive stencil taken from the same photographic image. The image was then printed in process colors, the screen carrying the negative stencil being printed first in yellow and then overprinted in magenta--but this time slightly out of register. The second screen, carrying the positive stencil, was then overprinted using cyan. This screen was again overprinted slightly out of register to produce this diffuse halftone version of the original source image. (Print designed and printed by Lal Sardar, aged eighteen, Gosford Hill School, Kidlington.)

Basic Proof Registration and Printing Technique

The printing of multicolored designs demands an accurate method of synchronizing a series of screens with the print paper.

Registration is quickly achieved by centering and then taping the original artwork on a sample sheet of print paper intended for use.

1

2

This is then positioned under the lowered screen so that the artwork corresponds exactly to its stenciled counterpart on the mesh.

3

Next, two side edges of the print paper are carefully bounded with cardboard tabs, each glued down to the printing bed. After removing the mounted artwork, insert a fresh sheet of print paper before printing begins.

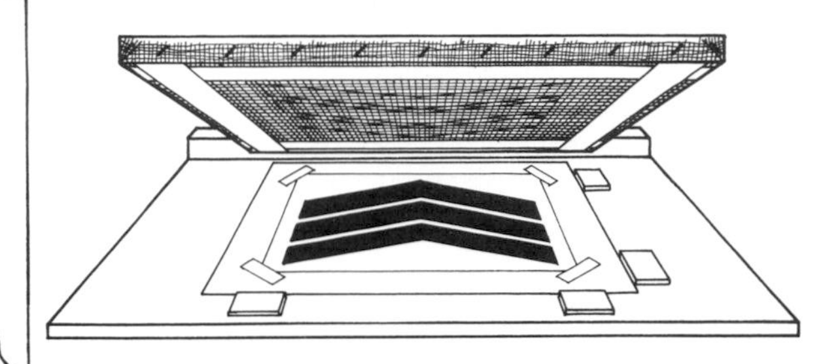

N.B.: Some screenprinters simply use strips of masking tape for registration. Only a slight upstand is required for locating the exact position of the print paper.

4

After the required color of printing ink is thoroughly mixed, the printing sequence begins. Ink is then poured evenly and generously along the "trough" of tape at the narrow end of the screen.

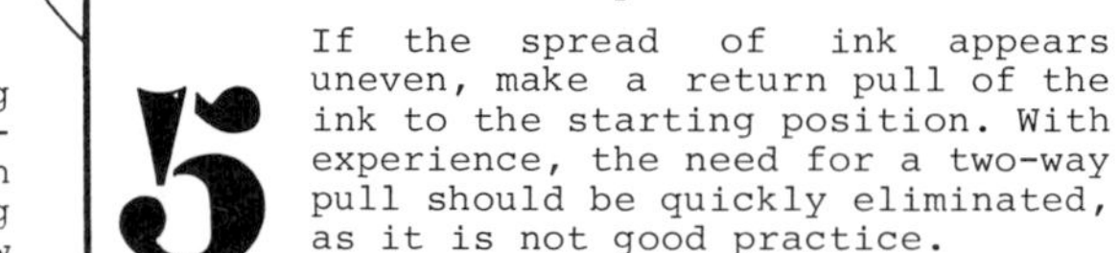

5

Next, place the squeegee behind the line of ink and draw it evenly but firmly across the screen, holding it at 45 degrees to the direction of pull.

If the spread of ink appears uneven, make a return pull of the ink to the starting position. With experience, the need for a two-way pull should be quickly eliminated, as it is not good practice.

6

Finally, lift the screen to reveal the proof. For long print runs the screw attachment of a small wooden swing prop to one side of the frame will keep hands free during proof removal and subsequent registration.

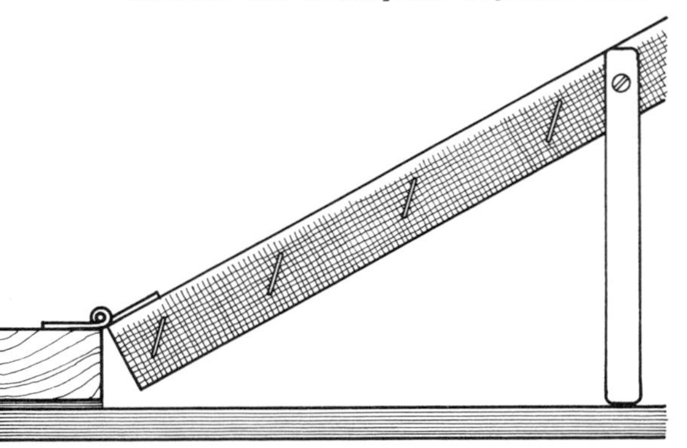

N.B.: To facilitate smudge-free drying, clip wet proofs to a line strung across the studio.

Screen Reclamation

1

The economics of screenprinting make it advisable to reclaim each screen after use. Therefore, after pulling the last print in a sequence, remove any surplus ink from the trough with a palette knife and, if the ink is uncontaminated, return it to its container.

2

Then place the screen face down on sheets of clean, absorbent paper, such as clean newsprint. Using turpentine and a clean squeegee, make a series of pulls to press the remaining ink through the mesh, then clean the screen with a rag.

N.B.: Alternatively, all traces of ink can be removed by washing the screen with a rag soaked in a proprietary ink solvent.

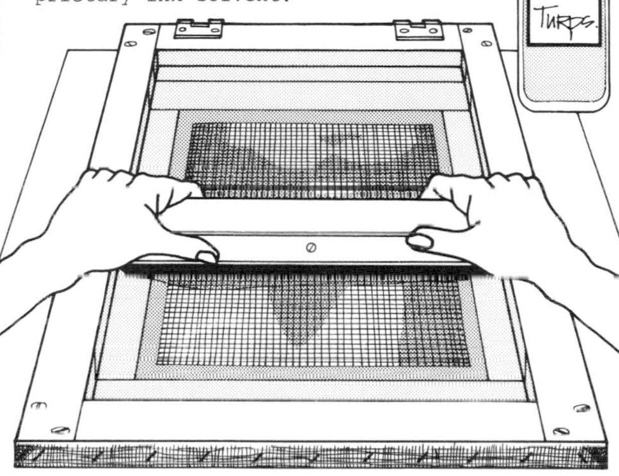

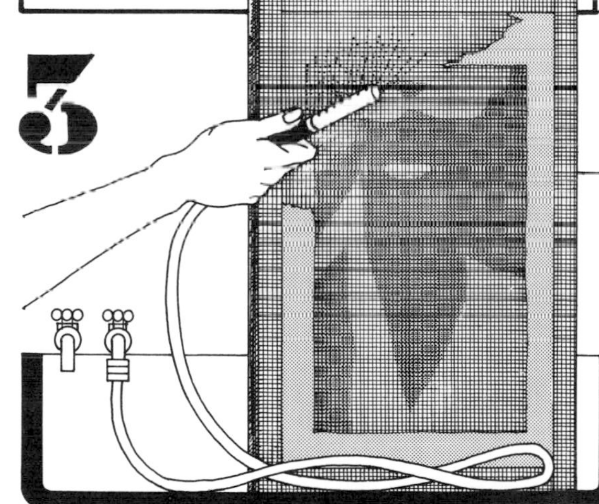

3

If a screen-filler medium has been used, this should now be removed before applying stencil decoating chemicals and to avoid mesh blockage. This is generally done by soaking the screen in water for a short period before hosing it down with cold water.

The next stage involves removal of the stencil. Each stencil type has an appropriate reclaiming product. For example, knife-cut paper and film stencils are removed after rinsing the screen with water and then applying a custom stripper to both sides of the screen, although scrubbing with a nail brush and warm water will do just as well. Emulsion stencils are removed by applying a compatible stripping medium and allowing this to act for the required time period.

N.B.: When using custom decoating agents, it is important to follow the product information carefully, making sure that cleansers and solvents do not dry on the screen before the washing-off stage.

5

Finally, remove the stencil by hosing it down with water from a shower spray attachment or, preferably, using a high-pressure water gun. With old or stubborn screens, decoating may not be entirely accomplished at this stage. In such cases, wash off the stripping medium and reapply the solvent to the still-damp stencil, waiting a few minutes before repeating the washing-off stage.

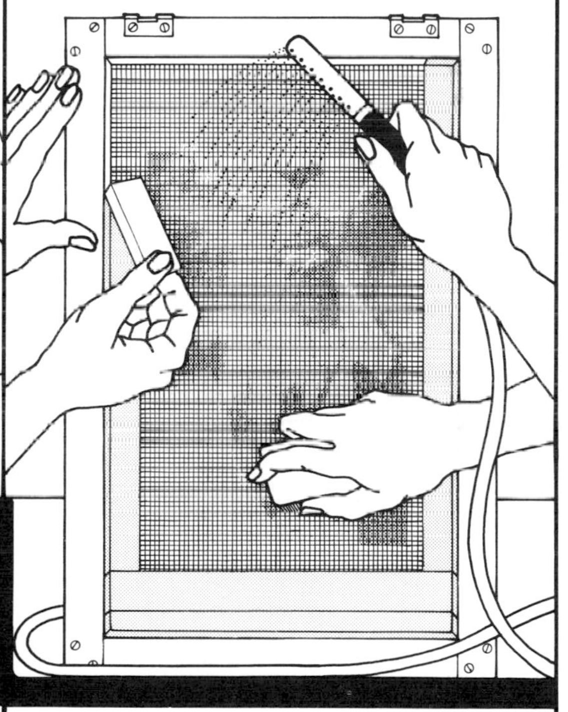

N.B.: It is good practice to remove all tape at this stage. Also, blocked areas of mesh can be opened by scrubbing them with a nail brush and water.

How to Make Marbling Effects

1

Marbling is one of the simplest methods of "printing" decorative color designs--bright, lively patterns being "lifted" from colors floating on the surface of a size solution. For this purpose, artists' oil paints provide a rich color range that, being antithetic with water, offer an exciting introduction to this technique.

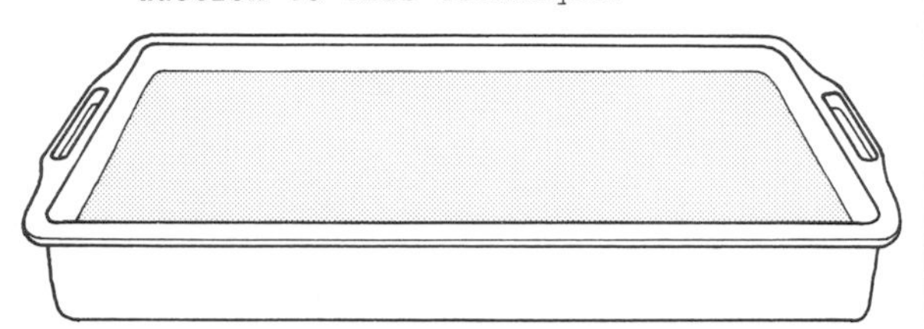

First, fill a baking tin or a photographic dish three-quarters full with a solution made from one filled dessert spoon of powdered gelatin size and one pint (0.5 liter) of water. To mix, dissolve the gelatin by adding a little boiling water before topping off with cold water. Allow the solution to cool to room temperature.

2

Select a range of oil colors, and squeeze about 1" (12mm) from each tube into small bottles. Dilute each with turpentine until runny.

N.B.: Keep each color well stirred, using a series of mixing sticks.

3

Next, drip a few drops of each color onto the surface of the solution, using the mixing sticks. Aim for regular intervals about 2-3" (50-75mm) apart.

N.B.: Each color should float on the surface and expand to a blob about 1 1/2" (38mm) in diameter. If the color does not spread, it needs further dilution. If the color sinks, the size needs further dilution. If the blobs expand or contract, this indicates that the size is either warmer or cooler than room temperature.

4

Once the colors have been floated, they can be arranged and rearranged by controlled agitation with the point of a knitting needle until the desired pattern is achieved.

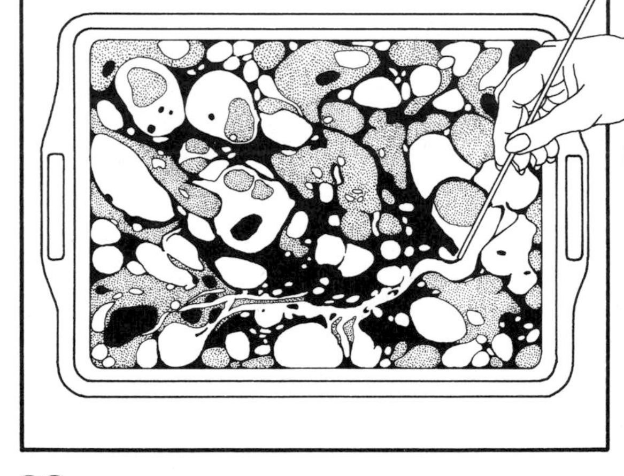

5

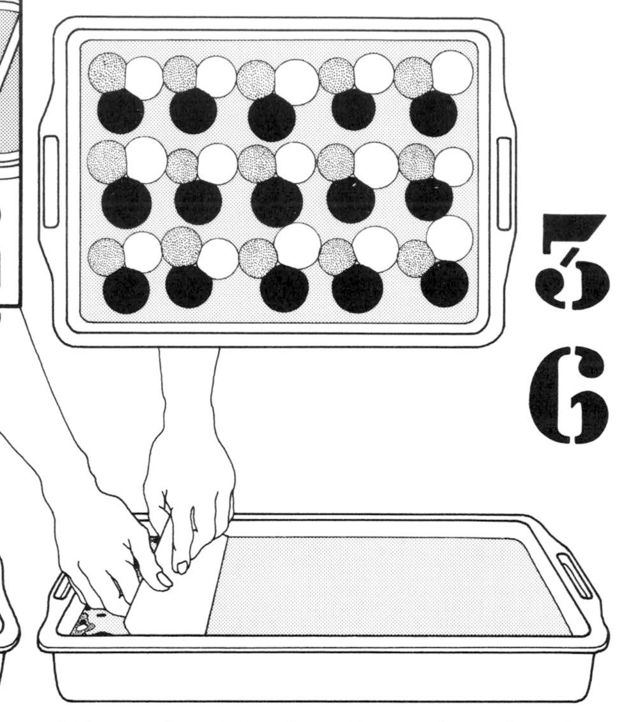

Then carefully lower a sheet of white drawing paper--with dimensions slightly less than those of the dish--onto the surface of the solution. Avoid air bubbles by starting at one corner.

When the entire sheet has been in contact with the color, lift off with equal care. Then hold it face down over the dish to drain before placing it face up on clean paper to dry.

6

After each print, skim the waste color from the solution by "combing" its surface with a strip of newsprint. Repeat the marbling process after reintroducing further colors.

N.B.: Marbled patterns will vary according to the amount of paint used and the order in which it is introduced to the solution. When fully dry, prints can be pressed flat with a warm iron.

Lithography: Basic Rendering Mediums

1

It is the granular quality of the drawing surface that makes the lithographic medium characteristically quite different from those described thus far. For example, a whole range of subtle values, lines, solids, and transferred textures can be incorporated into one printed design. These can be made directly and spontaneously onto metal or stone plates as one would work on textured drawing paper. As with the stones, lithographic metal plates are grained for suitability to crayon and ink. The various grades of crayon and ink (the latter sometimes called "tusche") comprise different proportions of greasy ingredients such as wax, soap, tallow, shellac, and lampblack. The lithographic printing process relies upon the natural principles of the mutual repulsion of grease and water and the chemical action of gum arabic. The following pages describe the rudiments of this process.

2

The lithographic drawing medium is litho crayon, a square sectioned, waxy black stick that is available in five density grades and also in wood-encased pencil form. The crayon can be pointed for fine line work or applied on its side edge or side corner, for rendering broad areas of tone. To avoid breaking the crayons, sharpen them by cutting backward carefully from the point.

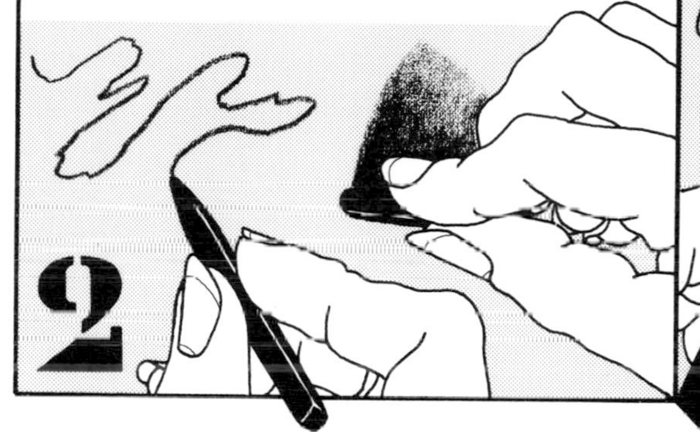

3

Other dry mediums that will register on the plate are lead pencils, Chinagraph pencils, oil pastels, and ballpoint pens. However, their image-producing capacities should be tested by the beginner before making any widespread use of them on a plate.

4

A liquid litho medium is prepared by rubbing a special stick ink into a small quantity of distilled water that has been heated in a tin lid over a gentle flame. Although various densities of wash can, with experience, be exploited, a reliable printed image results when a wash or a line application dries to a solid black on the plate.

5

Applicators for the liquid medium range from sable brushes for lines and washes, stencil brushes for stippling and splattering effects, and found surfaces for printing textures, to dip pens for extra-fine line work.

N.B.: Remember that the image on the plate is seen against gray zinc, whereas the ultimate print may appear as black or in color on white paper. It may be advisable therefore to render a little more emphatically than the desired result.

6

A noteworthy method of protecting the sensitized plate from unwanted grease marks during rendering is to apply a thin film of gum arabic to the areas, such as the frame around the image, that will not receive ink. This is applied using a clean sponge after the key guideline drawing has been transferred to the plate (see page 91).

N.B.: If registration marks for color printing are to be included, these should be ink rendered before the gum is applied (see page 95).

Lithography: Experimental Ink Effects

1 It is wise for the beginner to produce a test plate on which a whole range of experimental "dry" and "wet" ink techniques and effects are explored. For example, this is a detail of such a test plate made by Johanna Ashby, a student at the Ruskin School of Drawing and Fine Art, Oxford. It includes various intensities of litho crayon, dry-brush work, ink splattering, and textures achieved by dabbing the plate with various "found" applicators, such as crushed paper, etc.

2 Litho ink in liquid form can be exploited to make different wash effects. These can be created by using various dilutions, by mixing them with different mediums, or by employing unusual application techniques.

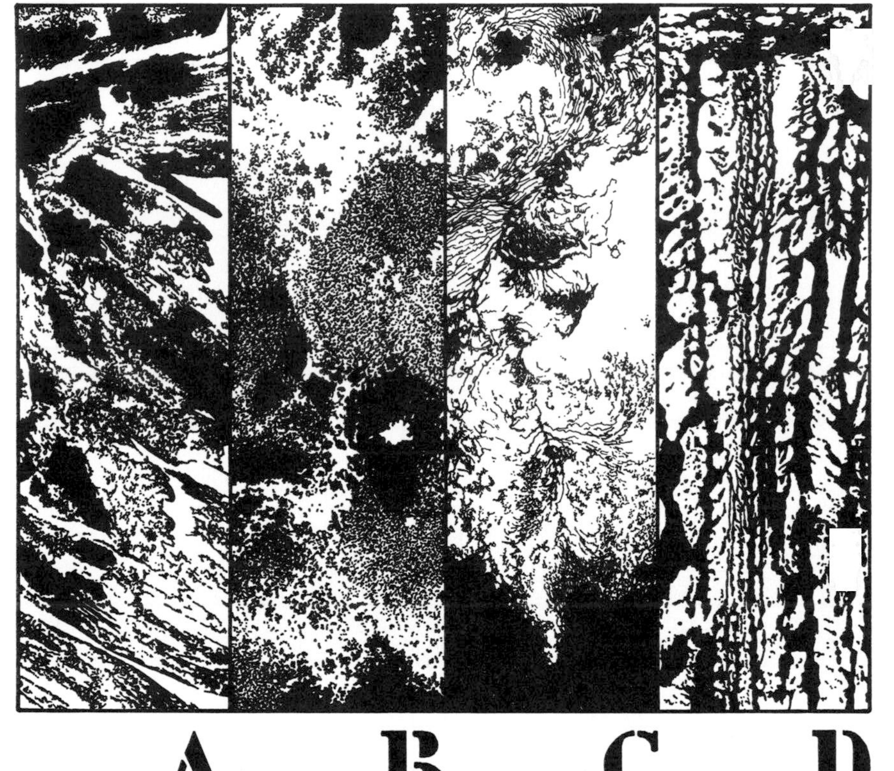

A B C D

A A textural wash created by dropping litho ink into pools of water previously applied on the plate.

B A textural wash produced by dissolving litho writing ink in turpentine before applying it to a predampened plate.

C A textural wash resulting from a mixture of litho writing ink and distilled water that is floated on a predampened plate.

D A textural wash produced by brushing a solution of litho writing ink onto a plate predampened with carbon tetrachloride.

N.B.: The edges of textural wash effects can be controlled by first applying gum arabic as a mask.

Transferring Artwork from Original to Litho Plate

When creating multicolored lithographic prints, a separate plate is required for each of its constituent colors. The territorial extent of these individual colors, and the parts they play in color mixing by overprinting, is traced into a key drawing.

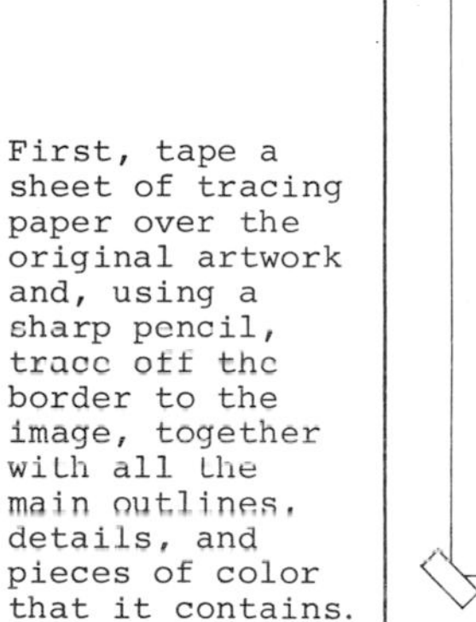

First, tape a sheet of tracing paper over the original artwork and, using a sharp pencil, trace off the border to the image, together with all the main outlines, details, and pieces of color that it contains.

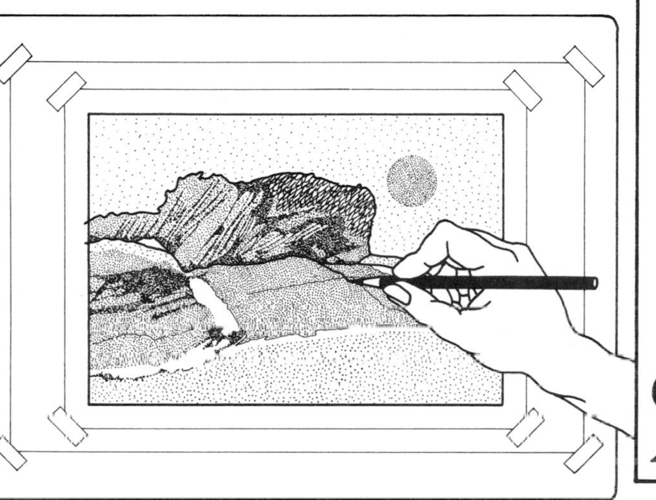

Outlines should be traced with a firm, accurate line and details with a finer, searching line. Extra-special care should be taken along the boundaries where two areas of similar color strength coexist, and also when recording the shape and extent of those pieces of color that are to be overprinted with others.

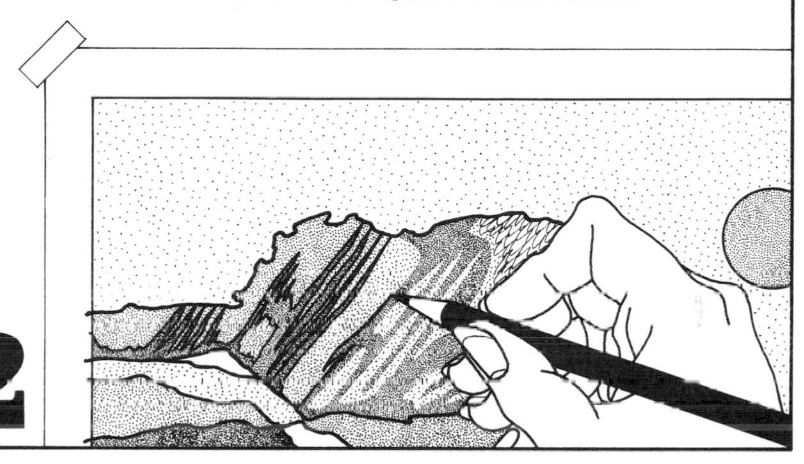

The next stage sees the careful transfer of all the traced information onto each of the plates in the color sequence. To achieve this, the master tracing is laterally reversed so that its image is mirrored, and placed over a sheet of lithographic transfer paper with its nonprinting, red chalk surface face down. These are then placed onto the first plate and, after the tracing paper is secured by taping at the top edge, the transfer to each plate in turn is begun by retracing with a fine ballpoint pen.

N.B.: Registration marks that align each plate with the proof paper during printing should also be transferred accurately to each plate at this stage (see page 95).

3

Once all the information has been transferred, with reference to the original artwork, each plate can now receive its lithographic crayon or ink rendering, its application being guided by the red chalk transfer lines. Also, the rendering on each plate will correspond to its role in the color printing sequence. For example, a plate intended to print a yellow will be rendered so that all the yellow components of the design--together with those areas using yellow in overprinting, such as to produce an orange when overprinted with red--are accurately recorded.

N.B.: Because the plate is sensitive to dirt and grease, it is vital to render with clean hands and equipment and to protect the more exposed areas of its surface with clean paper during this stage.

4

Processing and Proofing a Metal Plate

Between the image-making and proofing stages, the completed image on the plate needs to be processed so that its positive areas, when charged with printing ink, are receptive to the ink, while the negative areas are not.

First, place the metal plate complete with its artwork on a work-top covered with clean paper. Then dust over the image with French chalk or talcum powder on a clean pad. This thoroughly dries the plate and enables the gum (applied in the next stage) to desensitize the areas of the metal not occupied by the image.

1

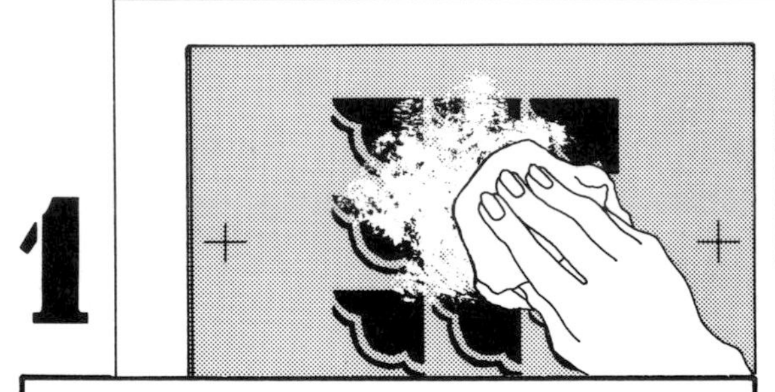

2

Next, wipe a thin film of gum arabic over the entire plate with a clean sponge. The gum will adhere to the areas of bare metal on the plate, but not to the greasy areas occupied by the image.

The gum is then dried thoroughly with the aid of a hairdrier or fan.

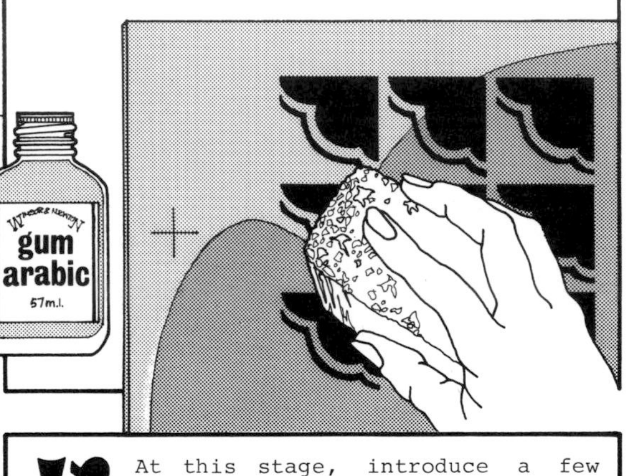

3

A do-it-yourself fan (traditionally used in lithography) is made by folding a piece of thin card around a wooden handle. The card is then stapled to form a sleeve, which provides a rapid drying action when swirled just above the plate.

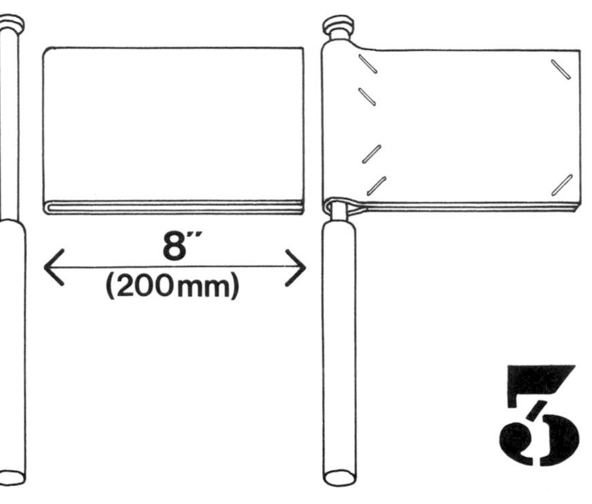

4

After making sure that the plate is dry, apply pure turpentine on a soft cloth to dissolve as much of the image as possible. Use a gentle circular motion to remove the deposits of black crayon and ink without removing the gum.

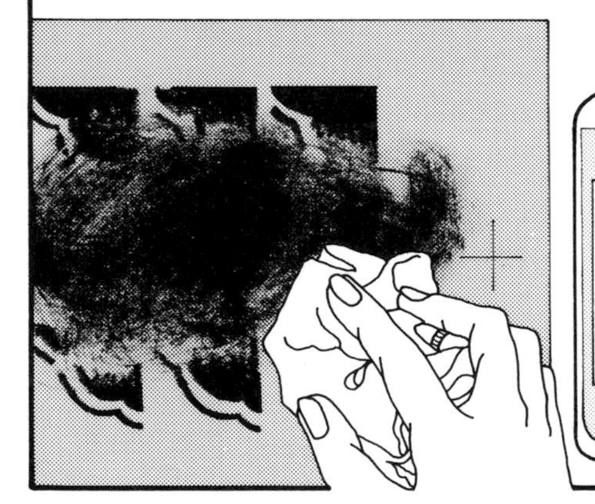

5

At this stage, introduce a few drops of liquid asphaltum onto the plate and gently wipe as an even film over its entire surface with a soft cloth. Allow this to dry before washing off the asphaltum--together with any remaining traces of turpentine or gum--with sponge and water.

N.B.: Asphaltum is a stop-out varnish, applied to protect the ink-receiving surface.

6

The image should now appear as a dry, discolored pattern against its moist background. It may be necessary to rub in asphaltum after the gum has been washed out. In this event, the plate must be flooded with water otherwise "scumming" of the non-image areas may occur.

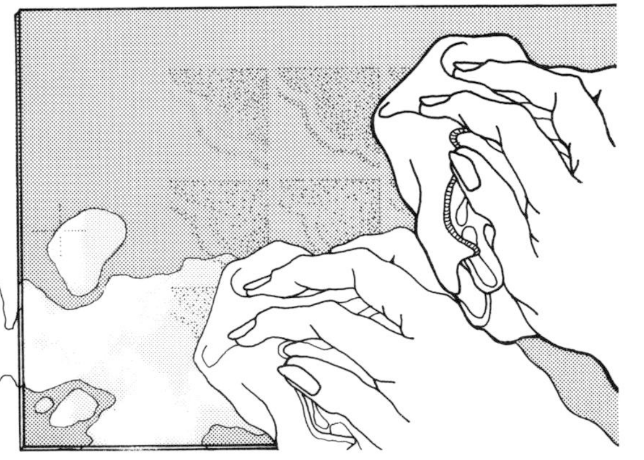

Processing and Proofing a Metal Plate

A proof can now be taken from the plate. To do this, the plate must be kept damp with a sponge.

7 Meanwhile, on a nearby litho stone or a plate-glass sheet, roll up some black proofing ink until it is spread evenly, with the roller fully coated.

8 Wipe over the plate with the damp sponge. Then roll up the image until it is fully charged with ink. After placing the plate into the press and overlaying the print paper, take the first proof. Repeat this operation until a solid black proof has been obtained.

9 The dampened plate should then be recharged with proofing ink and fan-dried before dusting over with French chalk or talcum powder applied with a clean cloth. Wipe away any surplus with a damp cloth.

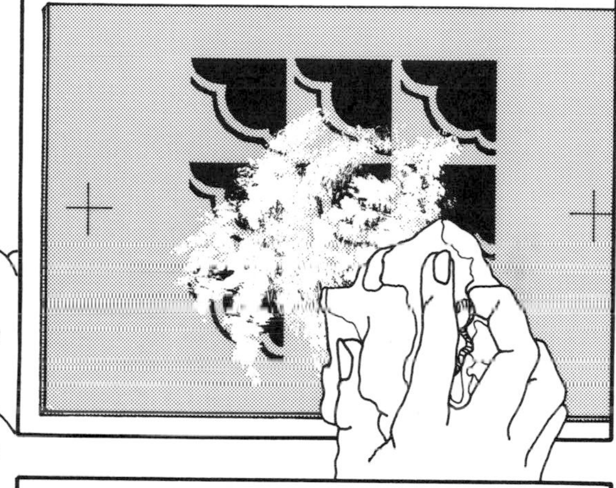

At this stage, any dirt marks or blemishes can be removed with a snakestone eraser stick while the plate is kept damp. Snakestones can be pointed by rubbing one end on a sheet of glasspaper.

N.B.: When using the snakestone, avoid damaging the plate by rubbing too hard. Also, make sure that your hands do not touch the plate.

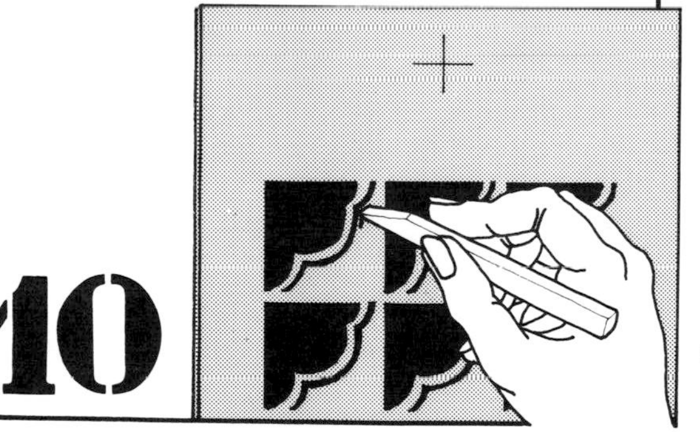

10

The plate is now ready for conditioning, by applying a proprietary etching solution appropriate to the kind of metal plate used. There are several "safe" etching mediums on the market that are diluted either in water or gum arabic before use. It is crucial that the manufacturer's instructions be followed, both for the manner of dilution and the duration of its contact with the plate.

Apply the plate etch with a small sponge, first coating the margin surrounding the artwork before recharging and coating the image itself. After leaving it for the required period of time, wash off the solution with sponge and water.

11

Finally, coat the plate liberally with gum arabic. Pour a pool at its center, then rub it in and spread with the palm of the hand. When thoroughly fan dried, the plate is ready for printing, or it can be stored.

N.B.: When required for printing, stages 4 and 5 have to be repeated before charging the plate with the required color of ink.

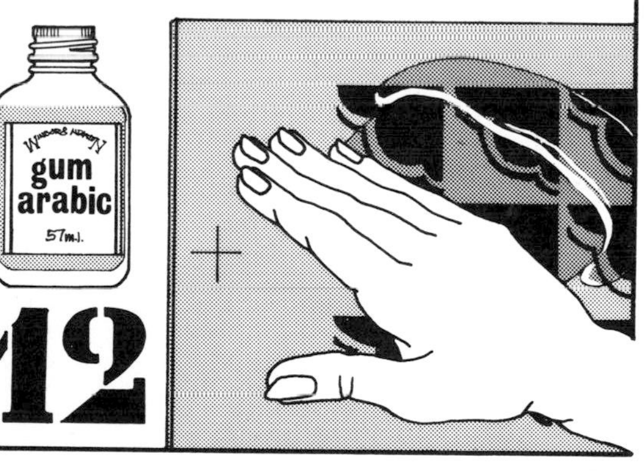

12

Printing the Plate After Processing

When the plate has been processed and you are ready for printing, place an adequate amount of print paper near the press. In order to keep the paper clean during printing, handle it with a folded cardboard strip.

Printing ink should next be rolled out on the ink slab. Make sure that the slab is perfectly clean, or bright colors can become contaminated. Always mix colors carefully, using a separate palette knife for each color.

1

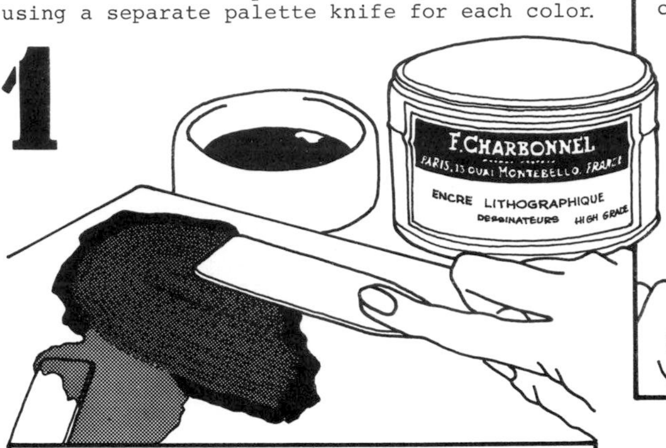

2

Once the ink has been thoroughly blended, deposit a line of ink at the far end of the slab. Then, rolling toward the ink, pick up and spread a fine, smooth film by rolling backward and forward.

N.B.: There are two kinds of rollers: nap rollers and rubber composition rollers.

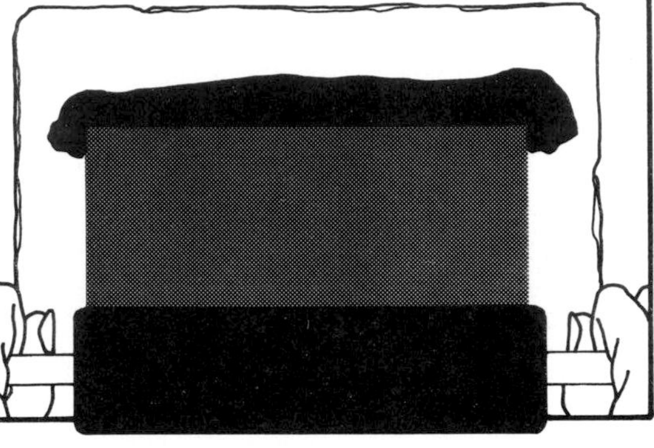

3

Next, place the plate in the bed of the press and wash out the image (see page 92, frames 4 and 5). Then dampen the plate with a clean sponge almost squeezed dry of water before rolling up the image in ink.

N.B.: If printing is to be in black or a dark color, the image should be washed out through the gum mask with turpentine, then wash-out solution. If the color is to be bright or light, only turpentine should be used. Then wash off the gum mask, damp sparsely, and roll up in color.

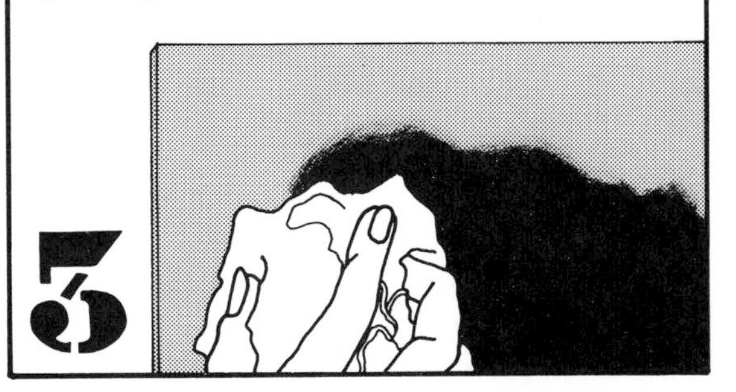

4

Finally, fan-dry the plate and take an initial proof on newsprint under a clean backing sheet. If the color is correct, moisten the plate again and roll up the image, using swift movements of the ink roller.

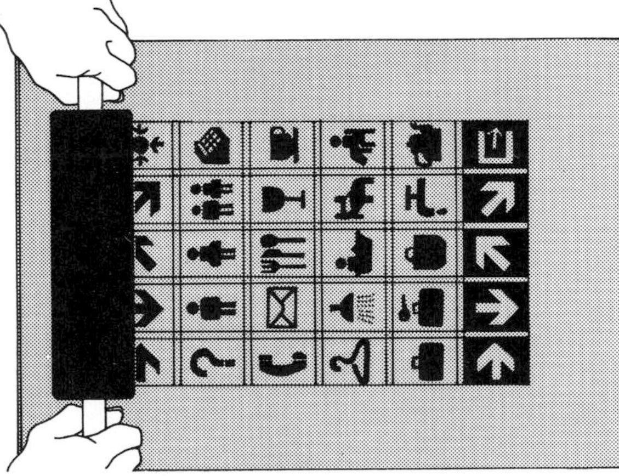

5

The old manually operated flatbed presses are ideal for short-run printing. The plate is mounted on a lithographic stone and, with its print paper and backing sheet, covered with a metal or leather tympan before being wound under pressure beneath a fixed leather covered wooden scraper.

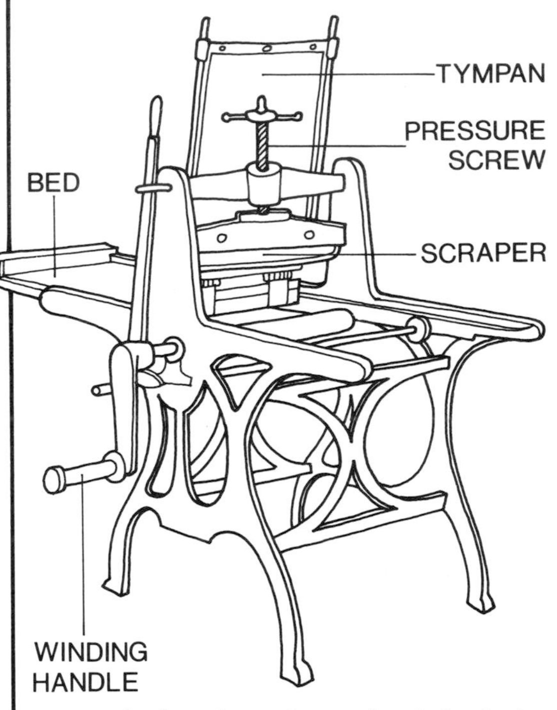

N.B.: If the plate is to be left during printing for any length of time, it should be gummed up with gum arabic. If a plate with an image is to be stored for later printing, it must be rolled up in black proofing ink, gummed, and stored in a dry place.

How to Register Color Proofs

1 The basic method of registering each plate to the proof paper when producing multicolored prints is to mark two small crosses at each end of the original artwork. These are located well outside the margin of paper required to frame the completed print.

2 Then, via the tracing from the original artwork, the crosses are transferred to (and processed on) each of the color plates in the print sequence.

3 Alignment between print paper and plate is achieved after proofing the first color by cutting a triangular window into one quadrant of each printed cross. When the proof is laid down on subsequent color plates, the paper is located so that each window coincides exactly with its former position over the crosses.

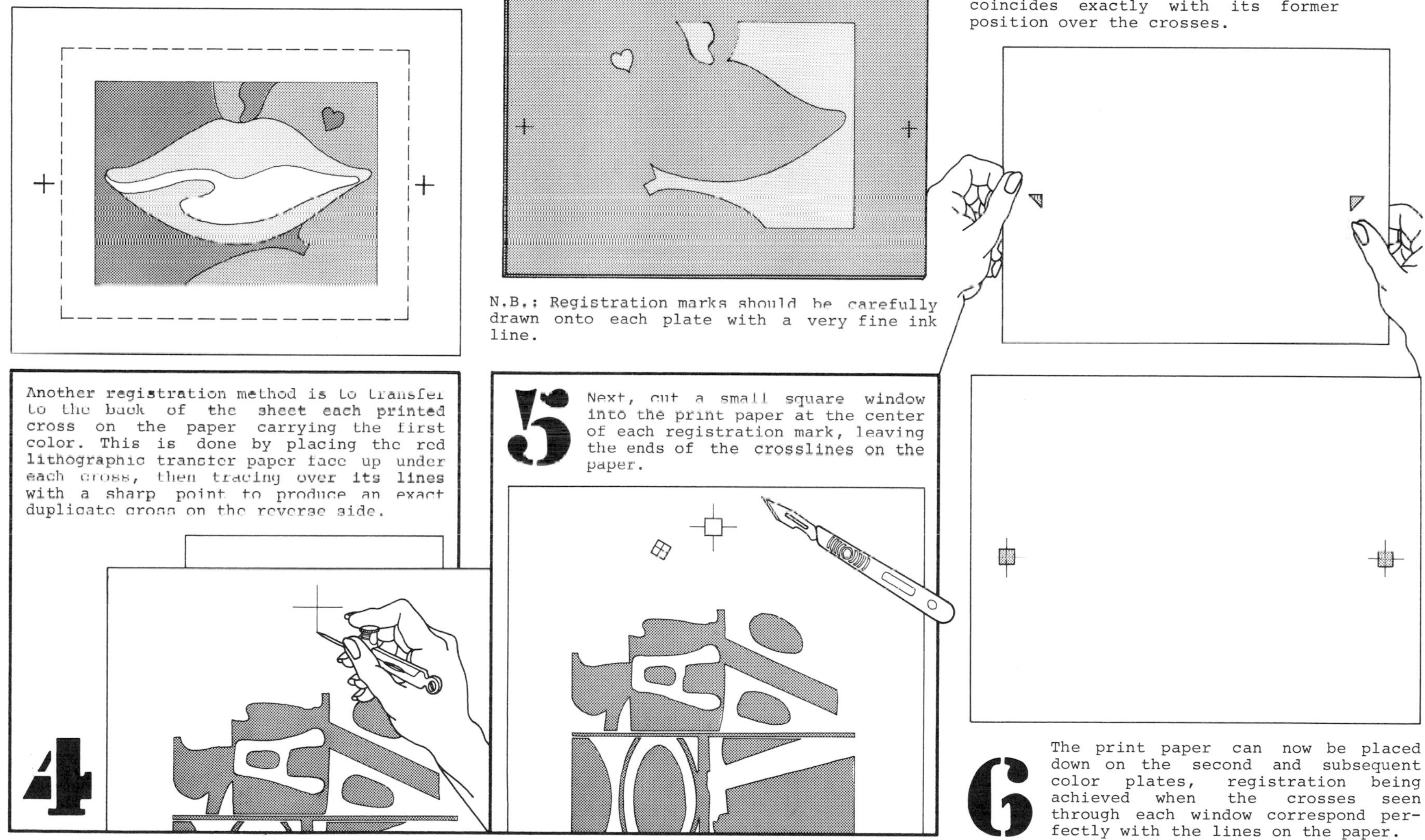

N.B.: Registration marks should be carefully drawn onto each plate with a very fine ink line.

4 Another registration method is to transfer to the back of the sheet each printed cross on the paper carrying the first color. This is done by placing the red lithographic transfer paper face up under each cross, then tracing over its lines with a sharp point to produce an exact duplicate cross on the reverse side.

5 Next, cut a small square window into the print paper at the center of each registration mark, leaving the ends of the crosslines on the paper.

6 The print paper can now be placed down on the second and subsequent color plates, registration being achieved when the crosses seen through each window correspond perfectly with the lines on the paper.

Lithograph-Screenprint Combination

This is a detail of a combination lithograph and screenprint of a pork chop made by Bridget Corkery, a First Year Graduate student at the Florida State University, Tallahassee. The lower section of the print contains textures made by pressing an actual port chop, that is, the subject of the drawing, directly onto the sensitized plate.

4 MODELMAKING TECHNIQUES

Functions of Models in Design

1

There is little doubt that solutions to spatial problems are best arrived at by working in three dimensions. For example, in an Oxford visualization test, groups of designers were asked to translate (without visual aids) ten verbally instructed and progressively more complex stages in the dissection of a cube. At the stage when unaided visualization broke down, subjects were given a pencil and paper and, again, when this was found inadequate, they were handed Plasticine and a knife. A comparison of the strike rates between unaided, graphic, and model sequences showed that use of a model enabled subjects to complete the entire sequence quickly and correctly in all cases. It was concluded that when graphics are the sole method employed in design, alternative solutions that might exist beyond their capacity could remain hidden.

2

At its most fundamental level, a three-dimensional model functions as a physical diagram. Its usefulness can be seen in everyday situations, such as in an after-dinner conversation, when found objects on the table are enlisted spontaneously to simulate and thus emphasize a topographical point in the discussion.

3

Similarly, in architectural design the conceptual model--comprising ready-made objects and junk materials--acts as a physical diagram of an idea. Used at the outset of design, it is a rapid means of simulating, albeit on a rudimentary level, the formal nature of a developing design.

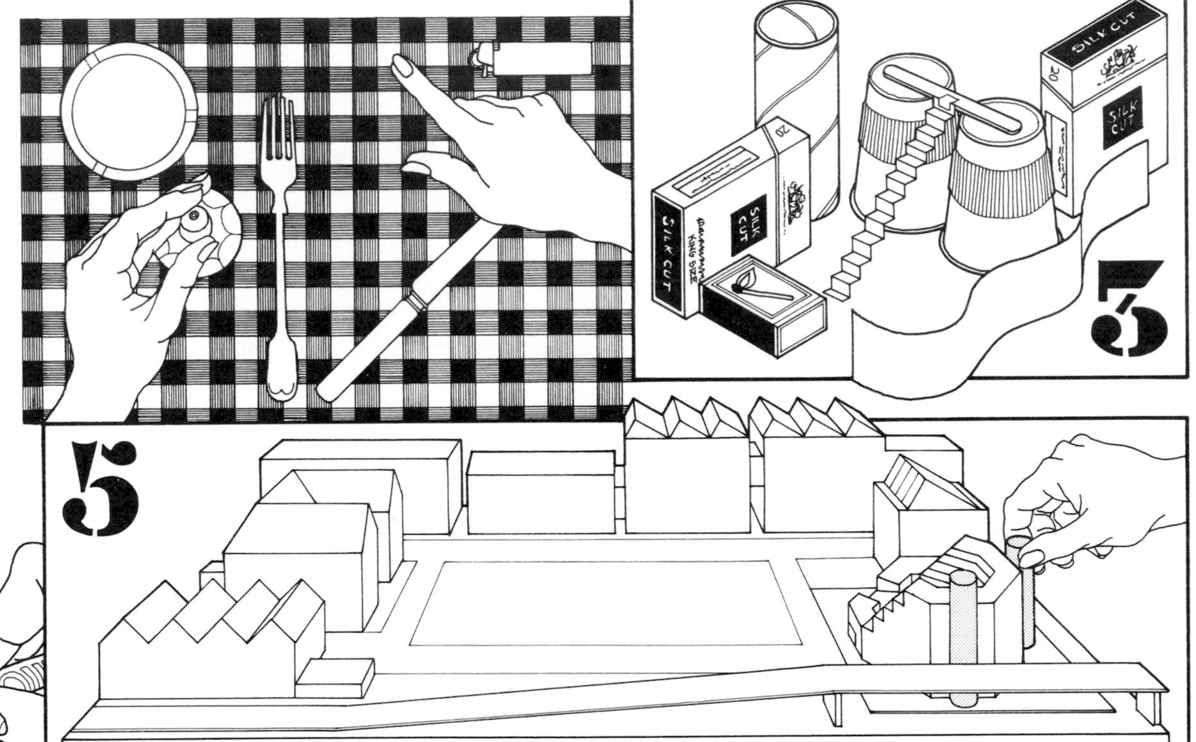

4

The sketch model is an important refinement of the conceptual diagram. It is an experimental instrument quickly constructed from scraps of paper and card, glue assembled to explore and check the various stages of an evolving idea.

N.B.: Because sketch models concentrate on the basic space-refining elements, they remain free from attention to surface detail.

5

The "fluidity" of sketch models is their salient feature, their response to changing ideas being quickly achieved using scissors or a scalpel. The formal implications of each modification to the design are then best tested against the context of a site model.

For this reason, site models should be constructed at the outset of a design project, this act in itself providing the designer with a deeper awareness of the building's setting. Site models are best built in block form, that is, simple carcasses in card that carve the external mass of existing buildings and objects that have bearing on a design proposal. In also providing the setting in which to communicate models of resolved schemes, site models retain their usefulness throughout the design sequence.

Functions of Models in Design

6

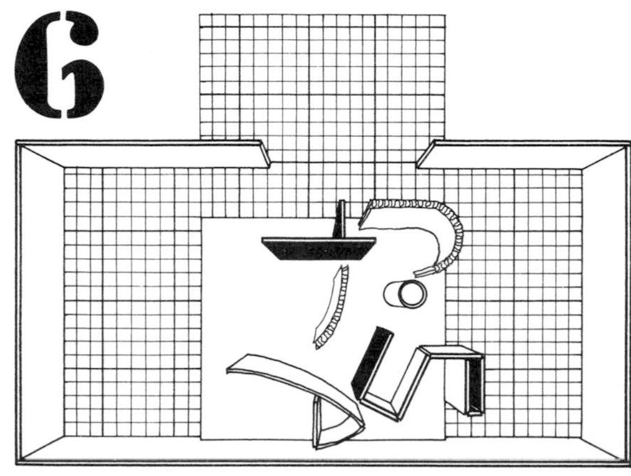

Another version of the sketch model is the space-planning maquette. This can take many forms but is used essentially as a popular design tool to make early excursions into the arrangement of more complex internal relationships, using folded or glued strips of paper or card.

7

As a design concept begins to take form, another type of model may enter the process. Although varying in precision and finish, the trial scale mockup is assembled to make in-depth studies of particular design aspects, such as the effects of atmospheric pressure, interior and exterior lighting, etc.

N.B.: Highly sophisticated versions are sometimes precision built to a large scale by professionals to study the efficiency of structures, acoustics, and the like.

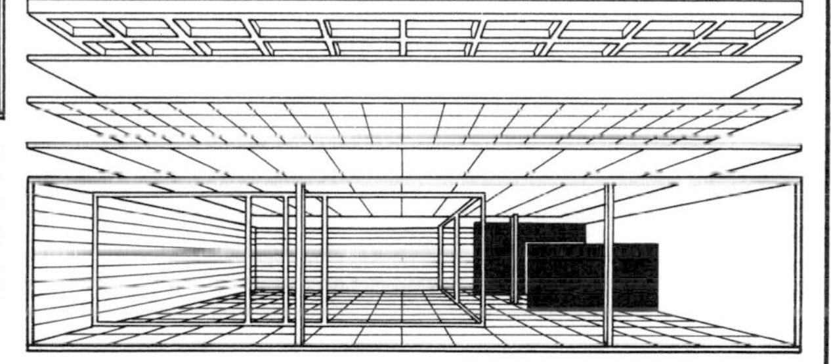

8

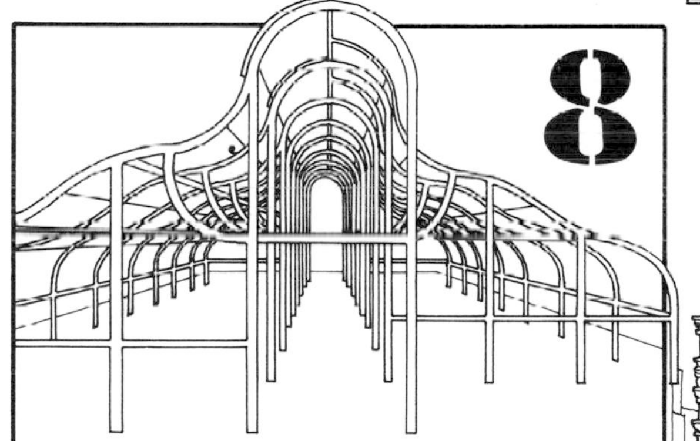

By examining the functional determinants of a building design isolated from surrounding features, construction models expose architectural private parts. They act as physical production drawings that study structure, assembly, or service systems, and recognize the important relationship between construction detailing and architectural form.

9

Presentation models represent the total composition of a design solution and communicate this finality to others. Being built primarily for promotion purposes, they can distort design intentions, if shown out of context or overinvested with meticulous detail.

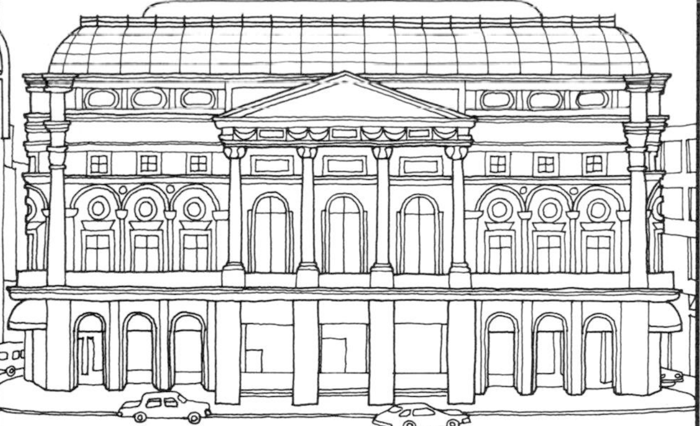

10

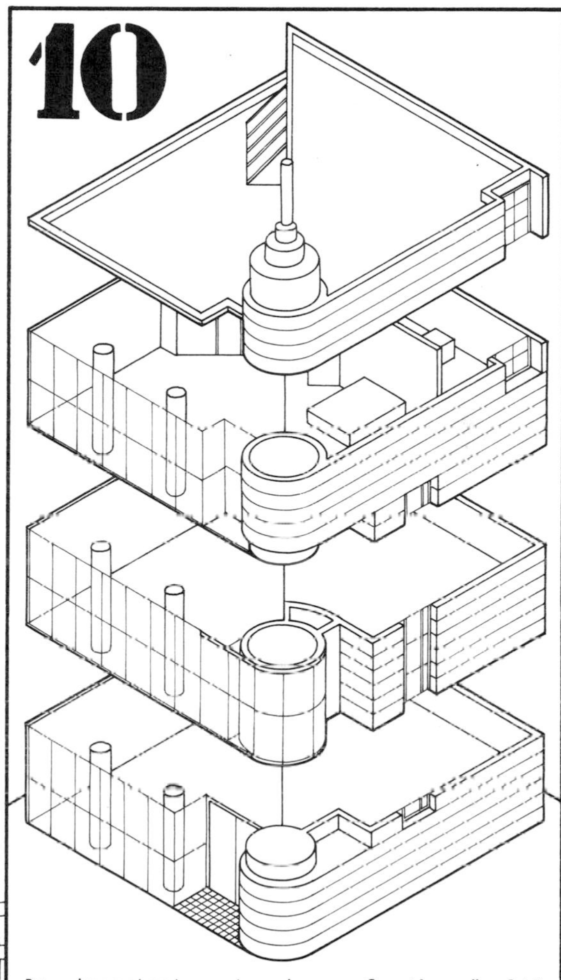

An important extension of the "solid state" presentation model is one that operates on many levels. For example, a demountable, multipurpose model can be preplanned so that its interiors can be viewed, its storys individually examined, and its construction details exposed--the complete assembly describing its external appearance.

Preplanning the Model

The selection of scale when making models is an important decision, as it relates directly to the type and degree of information that is intended to be imparted. Generally, scale models of individual domestic-sized buildings are constructed at 1/2" = 1' (1:20) or 1/4" = 1' (1:50), while the larger public, commercial, and industrial buildings and building complexes are manageable at 1/8" = 1' (1:100) or 1/16" = 1' (1:200). Extra-large scale developments and town-planning models of city sectors tend to be represented at a scale of 1:500 and even sometimes 1:1250.

This table serves to remind the beginner of the degree of detail afforded by these size representations and of how familiar forms appear and, indeed, disappear along the conventional modelmaking scales.

Another consideration concerns the basic method of construction, that is, whether models of buildings are to be produced as solid or as hollow forms.

Solid models of buildings exclusively describe external appearance and are quicker to construct. Forms are cut from wood or Styrofoam before being "dressed" with custom artwork or laminated with cutout design drawings.

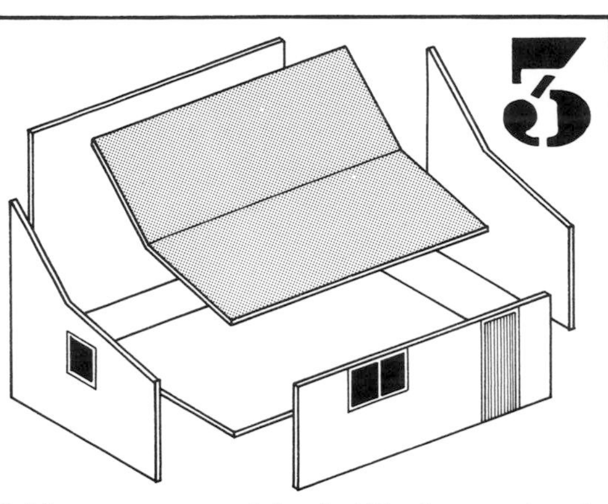

Hollow-carcass models built from cut-out and glue-assembled planes of cardboard are (if not prerendered with appropriate artwork before assembly) often used as supports on which custom artwork or design drawings are laminated.

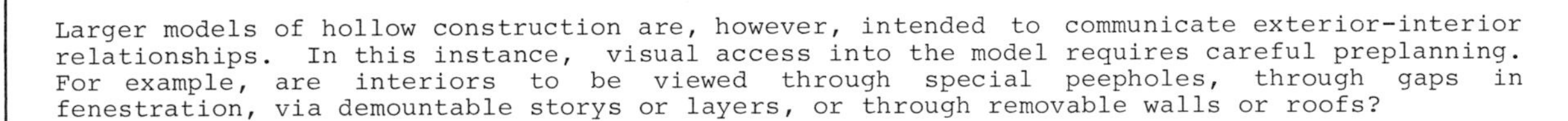

Larger models of hollow construction are, however, intended to communicate exterior-interior relationships. In this instance, visual access into the model requires careful preplanning. For example, are interiors to be viewed through special peepholes, through gaps in fenestration, via demountable storys or layers, or through removable walls or roofs?

N.B.: This decision, as all others, relies upon the nature of the design, your knowledge of it, and the kind of information you wish to convey.

Preplanning the Model

The preplanning of models also involves the selection of model-making materials and methods of color rendering. When time is precious and formal quality the message, achromatic models built from white card and relieved with steel-wool foliage are effective. Also, the combination of balsa wood, timbers, cork, glasspaper, and loofah provides an in-built harmony that can be heightened with the introduction of white card.

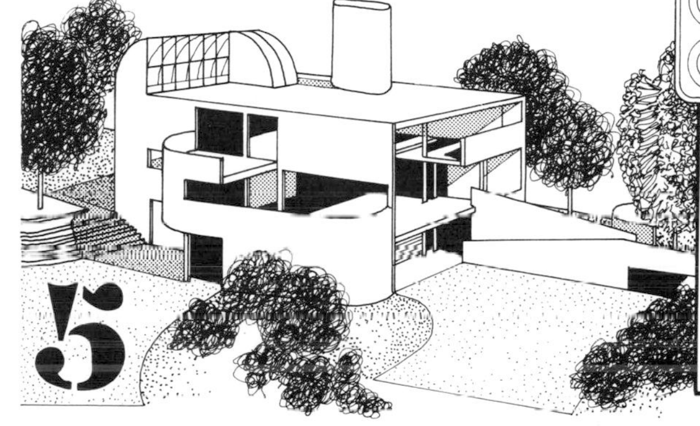

When color is applied to models, however, two basic functions operate: natural and schematic.

A decision to render natural color is a strategy that attempts to simulate the colors of building materials and natural substances as they would appear in the full-sized environment. This type of usage is usually reserved for presentation models.

A schematic color use, on the other hand, is an artificial color coding applied to emphasize the different aspects of a design, such as activity zones, service systems, and structural elements. Here color contrast is often exploited in models that present construction techniques or developmental models that diagram the mechanics of an idea.

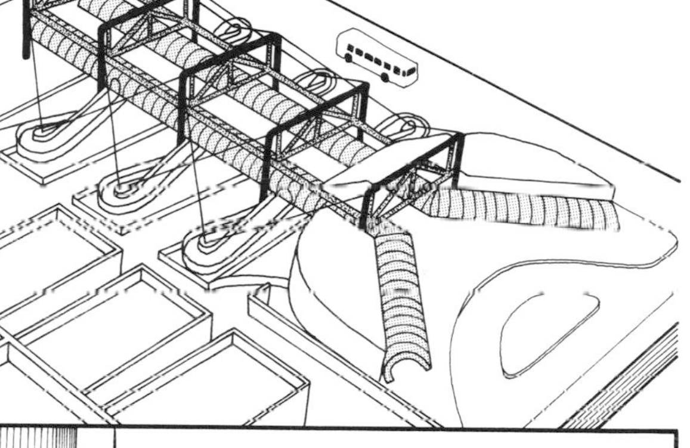

8 The decision that concerns the degree of detail or finish to be included in a model corresponds directly to the level of abstraction required for its role in design.

Although levels of abstraction tend to reduce along the design sequence, ultrarealistic models are to be avoided in presentation because higher degrees of detail can detract the viewer from the central, formal message.

It is important, however, to determine the degree of contrast between design and site models when they are presented in unison. For example, designers who are keen to show their proposal as a center of interest within the site context will overemphasize the contrast in treatment between the two. Others, eager to convey a sense of integration, will minimize or equalize this treatment.

N.B.: Being dependent upon whether the impact of a design proposal is local or more widespread environmentally, the territorial extent of existing information to be included on the baseboard of a site model tends to be self-determining. Make sure, however, that a site model can perform as a contextual setting for a design and not, if providing a restricted surround, merely act as an elaborate plinth.

How to Make a Battened Baseboard

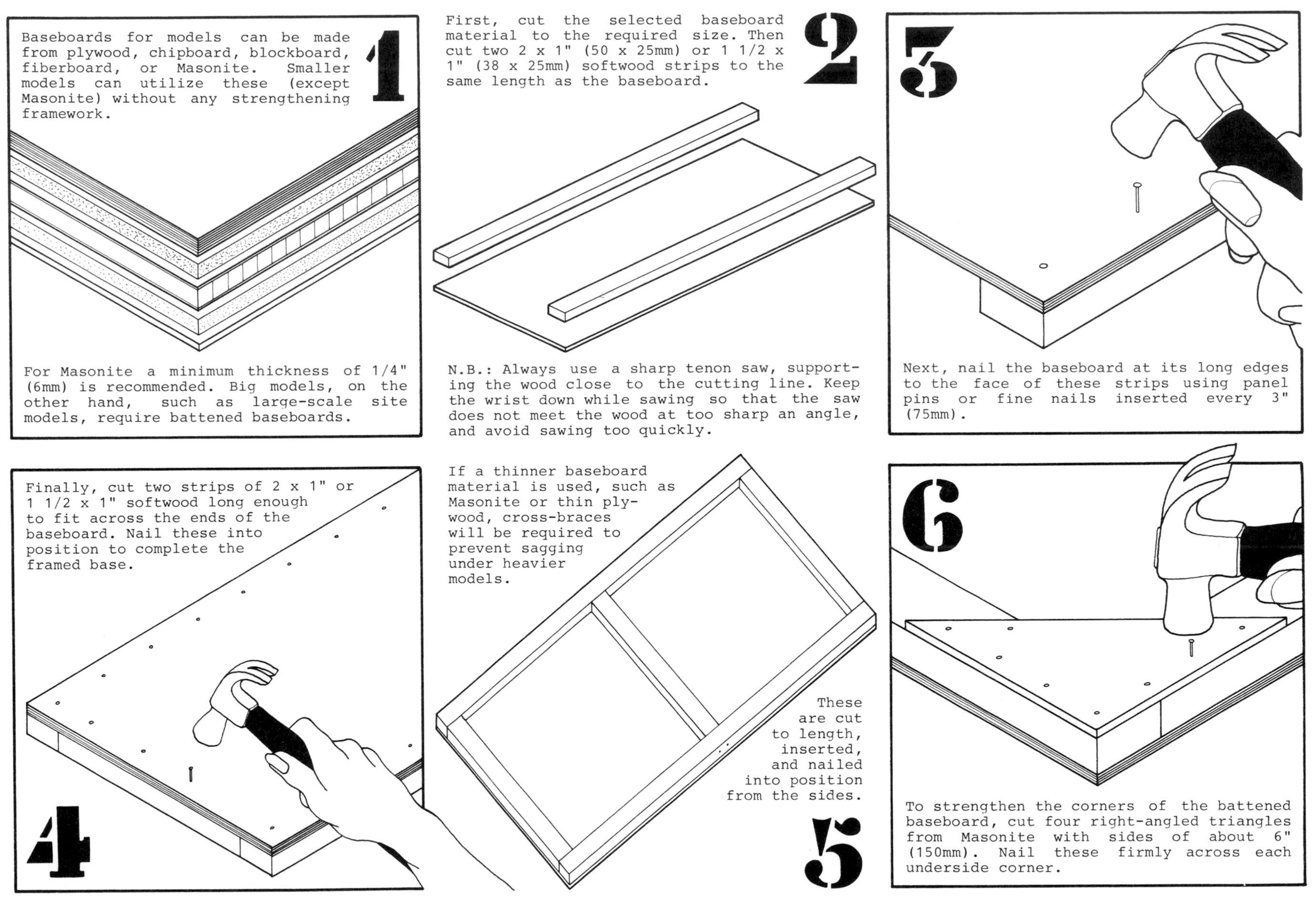

1 Baseboards for models can be made from plywood, chipboard, blockboard, fiberboard, or Masonite. Smaller models can utilize these (except Masonite) without any strengthening framework.

For Masonite a minimum thickness of 1/4" (6mm) is recommended. Big models, on the other hand, such as large-scale site models, require battened baseboards.

2 First, cut the selected baseboard material to the required size. Then cut two 2 x 1" (50 x 25mm) or 1 1/2 x 1" (38 x 25mm) softwood strips to the same length as the baseboard.

N.B.: Always use a sharp tenon saw, supporting the wood close to the cutting line. Keep the wrist down while sawing so that the saw does not meet the wood at too sharp an angle, and avoid sawing too quickly.

3 Next, nail the baseboard at its long edges to the face of these strips using panel pins or fine nails inserted every 3" (75mm).

4 Finally, cut two strips of 2 x 1" or 1 1/2 x 1" softwood long enough to fit across the ends of the baseboard. Nail these into position to complete the framed base.

5 If a thinner baseboard material is used, such as Masonite or thin plywood, cross-braces will be required to prevent sagging under heavier models.

These are cut to length, inserted, and nailed into position from the sides.

6 To strengthen the corners of the battened baseboard, cut four right-angled triangles from Masonite with sides of about 6" (150mm). Nail these firmly across each underside corner.

How to Make a Battened Baseboard

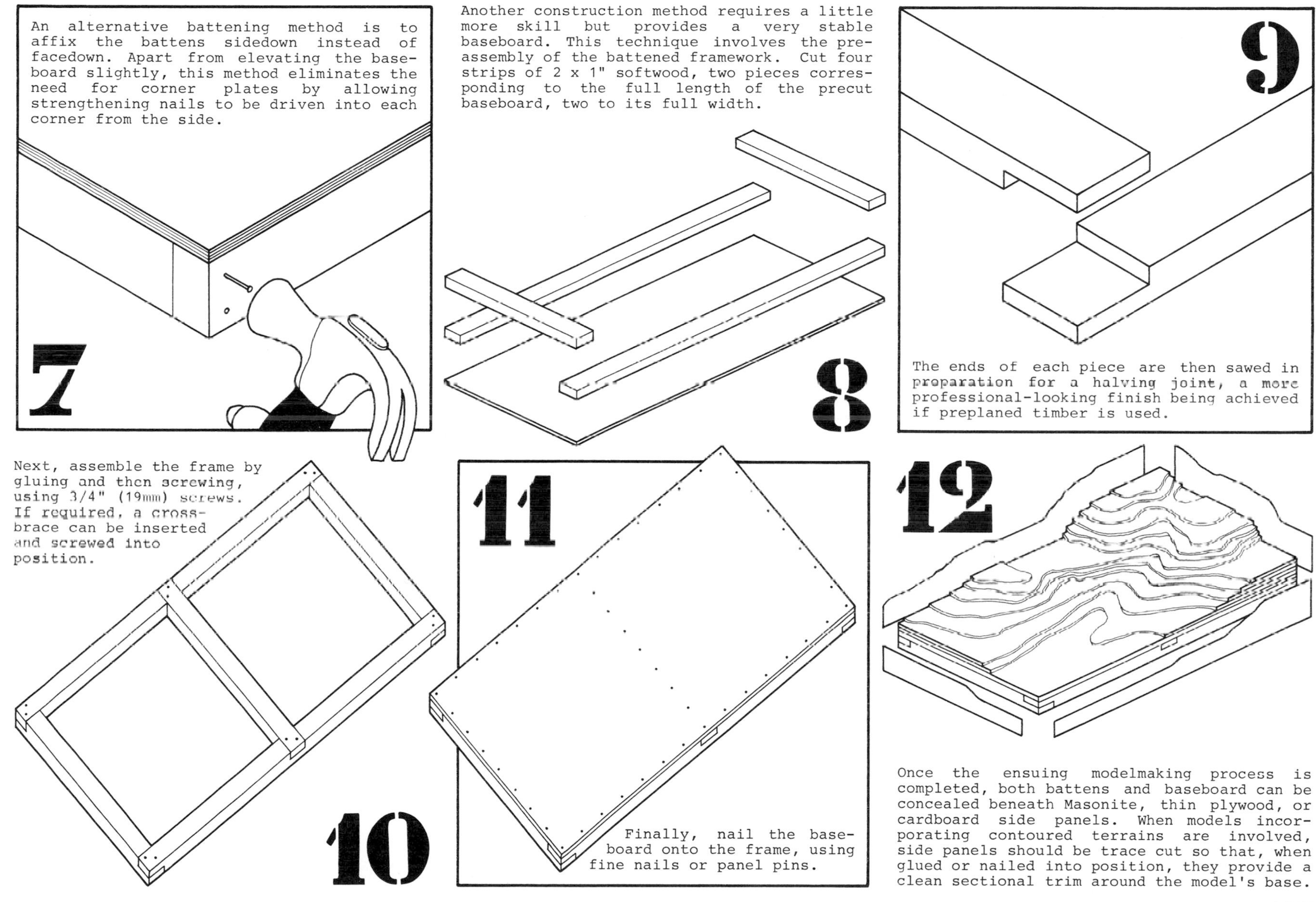

An alternative battening method is to affix the battens sidedown instead of facedown. Apart from elevating the baseboard slightly, this method eliminates the need for corner plates by allowing strengthening nails to be driven into each corner from the side.

Another construction method requires a little more skill but provides a very stable baseboard. This technique involves the preassembly of the battened framework. Cut four strips of 2 x 1" softwood, two pieces corresponding to the full length of the precut baseboard, two to its full width.

The ends of each piece are then sawed in preparation for a halving joint, a more professional-looking finish being achieved if preplaned timber is used.

Next, assemble the frame by gluing and then screwing, using 3/4" (19mm) screws. If required, a cross-brace can be inserted and screwed into position.

Finally, nail the baseboard onto the frame, using fine nails or panel pins.

Once the ensuing modelmaking process is completed, both battens and baseboard can be concealed beneath Masonite, thin plywood, or cardboard side panels. When models incorporating contoured terrains are involved, side panels should be trace cut so that, when glued or nailed into position, they provide a clean sectional trim around the model's base.

Modeling the Terrain: Laminated Contours

Ground heights are commonly modeled as laminated contours. These may be applied to solid or to reinforced baseboards (see pages 102-3).

Laminate materials include cardboard, balsa wood, Styrofoam, thin plywood, cork, Masonite, foam plastic, and, for prestigious presentation models, acrylic sheet.

1

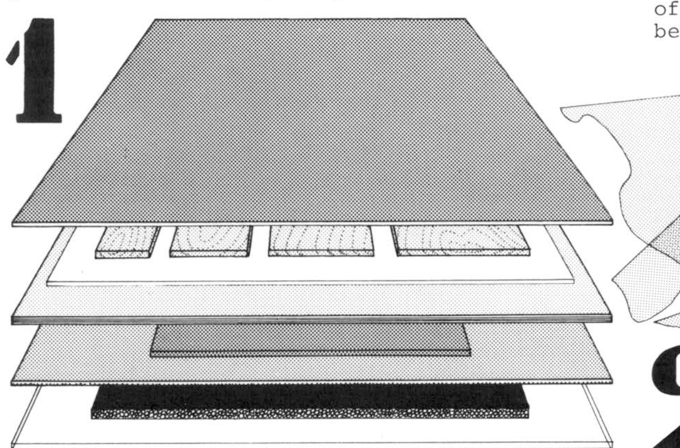

2

Contours are built up directly with reference to a site plan. Lamination begins with the lowest and therefore the largest area. This and subsequent shapes are transferred to the laminate material by tracing from the site plan with carbon paper.

N.B.: Some designers glue the cutout contours of a site plan directly to the laminate before reassembling a site plan in relief.

3

The contour is then cut from the sheet material selected for lamination.

N.B.: Cardboard, balsa wood, Styrofoam, and cork should be cut with a sharp, heavy-duty knife. Foam plastic can be cut with scissors. Masonite, plywood, and acrylic sheet can be cut manually with a fretsaw but are best cut using a power fretsaw or bandsaw.

4

After lightly sanding its edges with glasspaper, glue the contour (and nail it, if using timber or cork sheet) into position, making sure to use the appropriate adhesive.

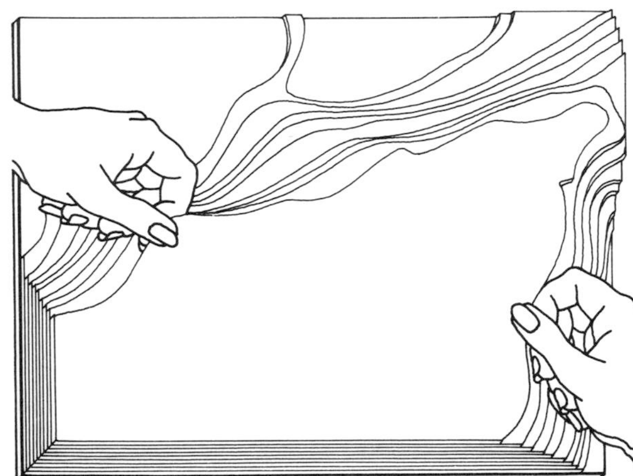

5

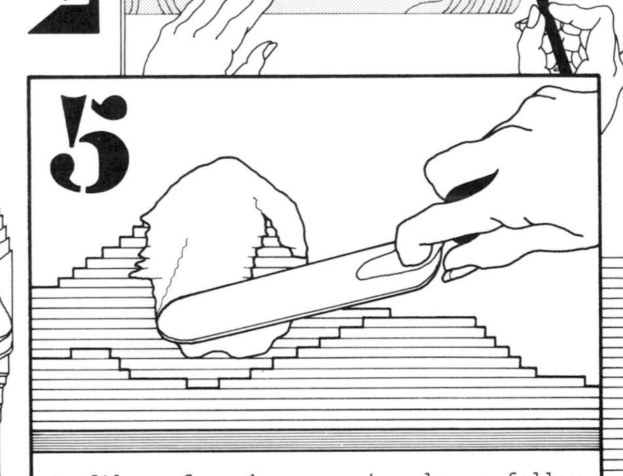

Profiles of each successive layer follow the appropriate plan contour line to complete the assembly sequence. If constructed in cardboard or wood, the finished terrain can be softened by an applied skin of water-based plaster crack filler, an elastic medium that can be sanded and painted when dry.

6

Excavations can now be made into the terrain. These may be required for the insertion of buildings, trees, sunken roads and rivers, etc., and can be made using knives, chisels, drills, and gouges.

N.B.: Excavations are not advisable if a hardboard or acrylic sheet has been used. In this event, excavations should be carefully preplanned and cut during the assembly stage.

Transferring Plan Contours into Models

The conversion of contours on a site plan into the scaled height of a lamination material for models of terrain is quickly achieved with a scale rule.

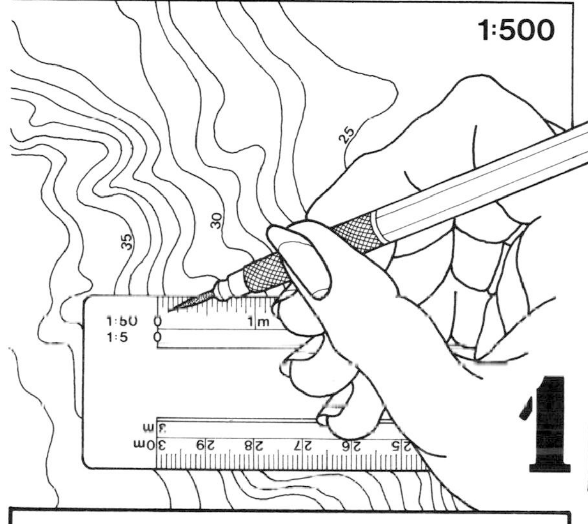

2

This table, however, shows the typical thicknesses of laminates for contour heights in relation to various modeling scales.

N.B.: When calculating overall thicknesses, always allow for the space taken up by the adhesive.

IMPERIAL

Model Scale	Height Between Contours-Feet	Lamination Thickness-Inches
1/4"-1'	1	1/4
1:48		
1/8"-1'	1	1/8
1:96		
1/16"-1'	1	1/16
1:192	2	1/8
1:500	5	approx 1/8
1:1250	10	approx 3/32
	25	approx 1/4
1:2500	25	approx 1/8
1:25000	125	approx 1/16
	250	approx 1/8

METRIC

Model Scale	Height Between Contours-Metres	Lamination Thickness-mm
1:50	0.1	2
	0.25	5
1:100	0.25	2.5
	0.5	5
1:200	0.5	2.5
	1	5
1:500	1	2
1:1250	2.5	2
	5	4
1:2500	5	2
	10	4
1:25000	50	2
	100	4

In models not demanding a high degree of detail, it is not always necessary to reconstruct every contour line on the plan. In such cases, alternate sequences of contours can be omitted and a thicker laminate material used. First, trace over the selected contours on the diazo print of the site plan . . .

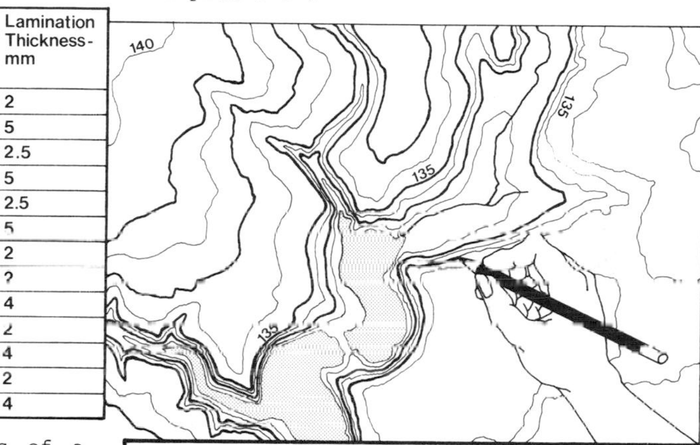

. . . or trace off from the original plan or map before transferring their shapes onto the model-making material.

N.B.: If one particular area is important to the site, select a characteristic contour and work upward and downward from this level.

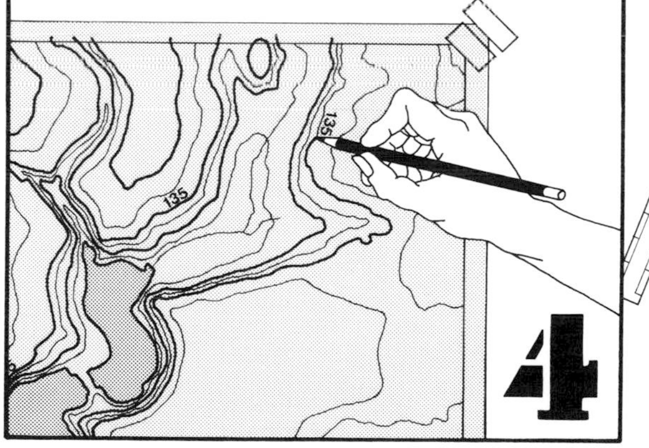

On the other hand, when the thickness of a laminate material falls halfway between two contour heights, it will be necessary to interpolate between them. At small scales this can be eyeballed, but where contours are widely spaced, an intermediate contour can be estimated by marking a series of center points between them.

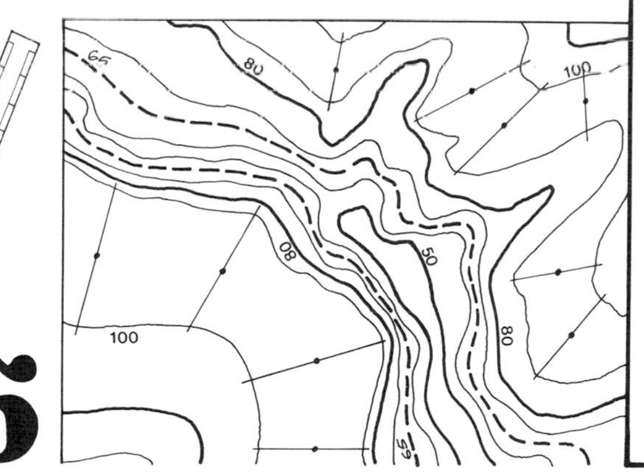

6

The points are then connected by a line that curves in response to the character of the two contour lines it separates.

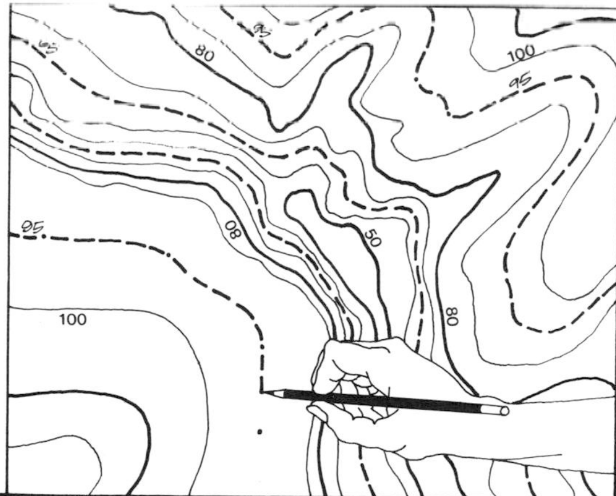

Modeling the Terrain: Reinforced Shells

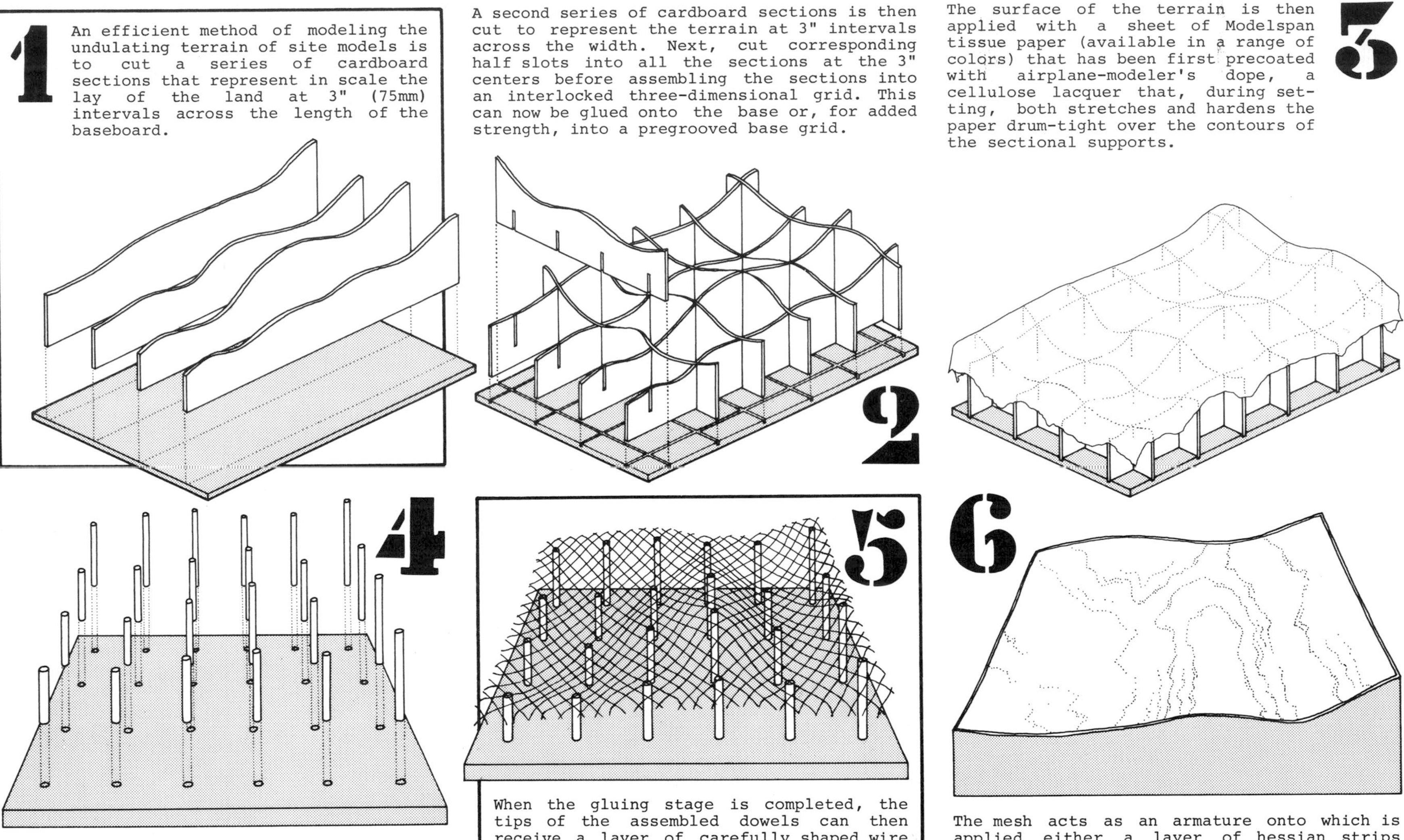

1 An efficient method of modeling the undulating terrain of site models is to cut a series of cardboard sections that represent in scale the lay of the land at 3" (75mm) intervals across the length of the baseboard.

A second series of cardboard sections is then cut to represent the terrain at 3" intervals across the width. Next, cut corresponding half slots into all the sections at the 3" centers before assembling the sections into an interlocked three-dimensional grid. This can now be glued onto the base or, for added strength, into a pregrooved base grid.

3 The surface of the terrain is then applied with a sheet of Modelspan tissue paper (available in a range of colors) that has been first precoated with airplane-modeler's dope, a cellulose lacquer that, during setting, both stretches and hardens the paper drum-tight over the contours of the sectional supports.

4 Another hollow-construction method is worked on a substantial wooden baseboard. After marking the board with a 3" grid, drill each center with holes of a regular depth. Into these glue a wooden dowel that has been precut to a length corresponding in scale to the equivalent spot height in the terrain.

5 When the gluing stage is completed, the tips of the assembled dowels can then receive a layer of carefully shaped wire mesh.

N.B.: The mesh can be further supported by a packing of paper or cloth before it is fixed, by tacking it onto the ends of the dowels.

6 The mesh acts as an armature onto which is applied either a layer of hessian strips dipped in plaster of Paris before a finishing coat of neat plaster, or a direct application of papier-mâché (see pages 110 and 111).

N.B.: The site model is completed after receiving its shaped side panels that are glued and nailed into position.

Transferring Plan Contours into Model Sections

1

First, superimpose the site plan with a grid drawn with 3" (75mm) centers.

N.B.: As the grid represents the location of sectional cuts, grid lines should each be numbered. To avoid confusion, this annotation is transferred at every stage in the process.

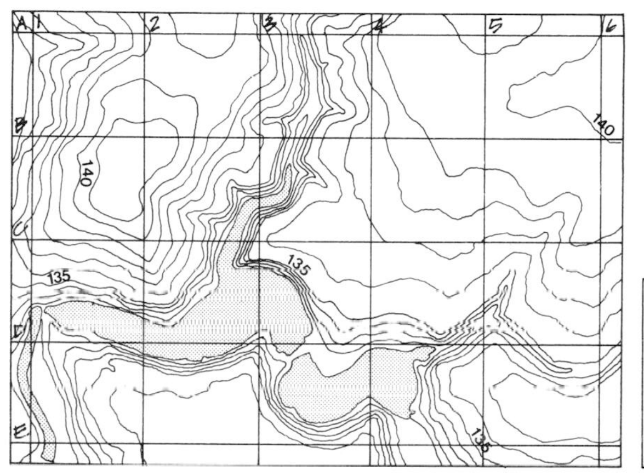

Next, position the edge of a strip of paper against each grid line in turn (a fresh strip for each grid line) and mark off the points and the heights at which the contour lines intersect.

N.B.: If a framework comprising cross- and longitudinal sections is required for the model, the points on all corresponding grid lines should be transferred.

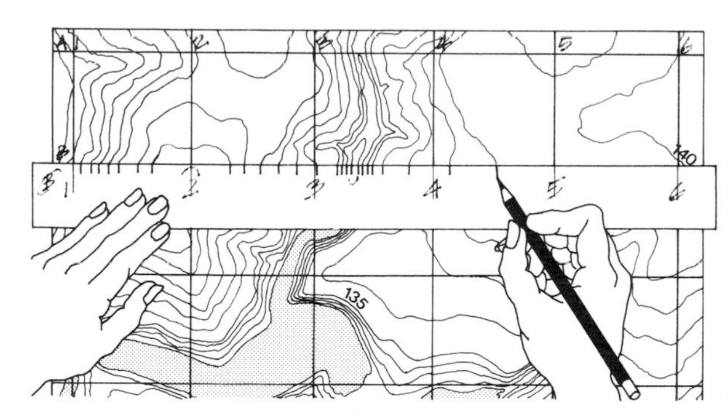

3

The next stage involves the transfer of each set of points on the paper strips--together with their annotations--to the model-making material.

To achieve this, simply project each set of points vertically through a series of scaled horizontal height lines drawn over a baseline on the material intended for the model.

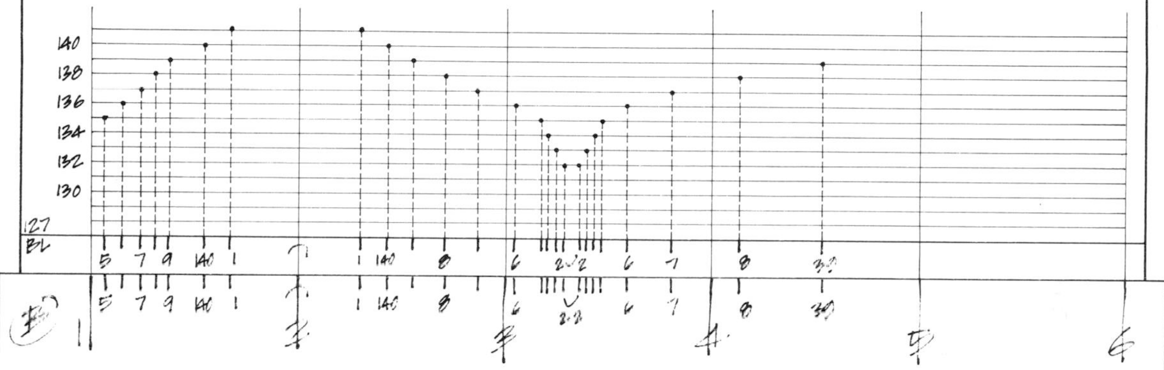

N.B.: The baseline is established first, located several contour thicknesses below the lowest contour level on the site plan.

4

Complete the drawing of the sections by connecting the transferred points with a smoothly curving line. Repeat this process for all the remaining sections.

5

Finally, cut out the shape of the contour sections and assemble as described on page 106.

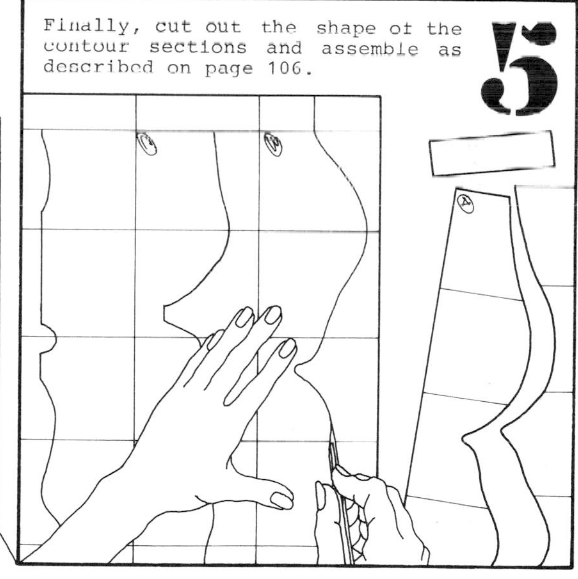

Modeling the Terrain: Sculptured Surfaces

1

A fast method of producing large areas of terrain is to use plaster of Paris. After mixing (see page 110), plaster is poured into a temporarily shuttered "bed" formed from strips of wood located around the edge of the baseboard, which are easily removed after the plaster has set. While the plaster is still fluid, however, its surface can be formed with the hands. When plaster is set, surfaces can be further refined with a chisel and, if required, smoothed with glasspaper.

N.B.: To avoid stress cracks, work the plaster models on rigid baseboards.

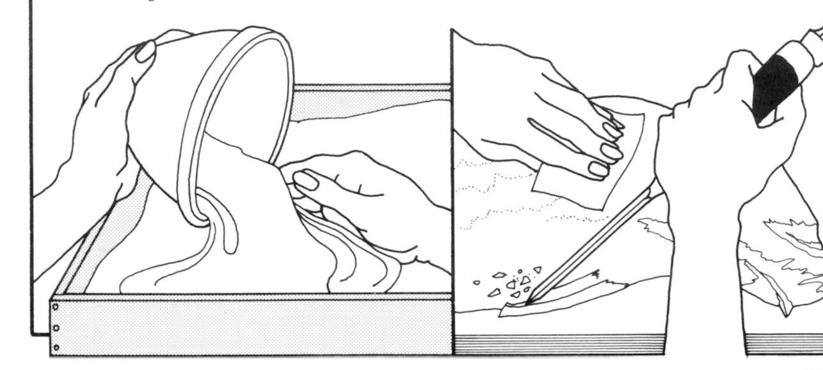

Another rapid method of producing large areas of terrain is to sculpture the surface of a block of Styrofoam, or several Styrofoam slabs glued together on a baseboard to provide the required size. The terrain is sculptured downward from its highest point by cutting with hacksaw blades or scalpels, or smelting with a thermosaw. A thermosaw is a hot-filament cutting tool especially made for working thermoplastics. However, a knife or wire heated over a flame will work just as well.

N.B.: Two important points to remember are, first, that proprietary nondissolving water-based pastes should be used when gluing Styrofoam. Other adhesives should be tested prior to use, as spirit-based glues decompose this material and are highly flammable. Second--and extremely important--when modeling Styrofoam with hot-wire cutters do not inhale the fumes, as they are highly toxic.

2

3

Small-scale models of terrain, especially of landscape designs, can be quickly modeled in Plasticine on a base of thick card or Masonite. A combination of working with the hands and a modeling spatula is ideal to achieve a rich variety of surface finishes, from glass smooth to highly textured. Also, as Plasticine is marketed in a basic color range, this versatility can be exploited, further colors being obtained by kneading two colors together.

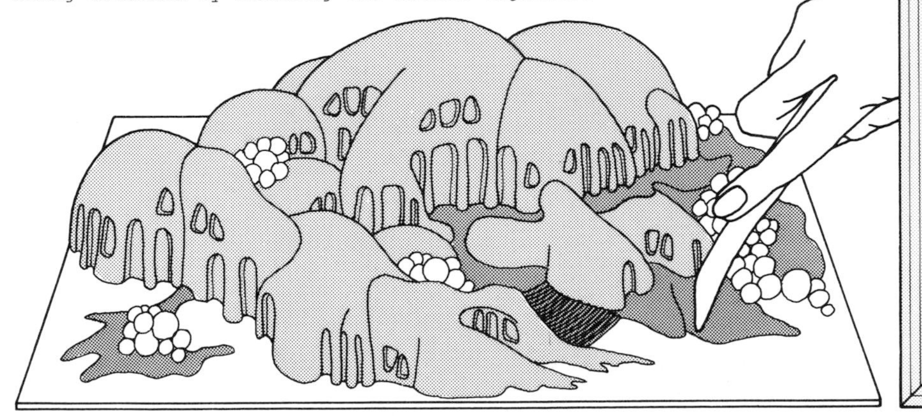

A block of wood comprising a glue-laminated sandwich of timber sheets can be carved with chisels and finished with power drills, files, and sanders. Although it requires no baseboard, this kind of terrain model takes time and a degree of skill. It is also most exclusively employed in client presentation, often being clear varnished rather than painted, to retain the natural colors of the wood.

N.B.: Nails should not be used in the construction of the laminated block.

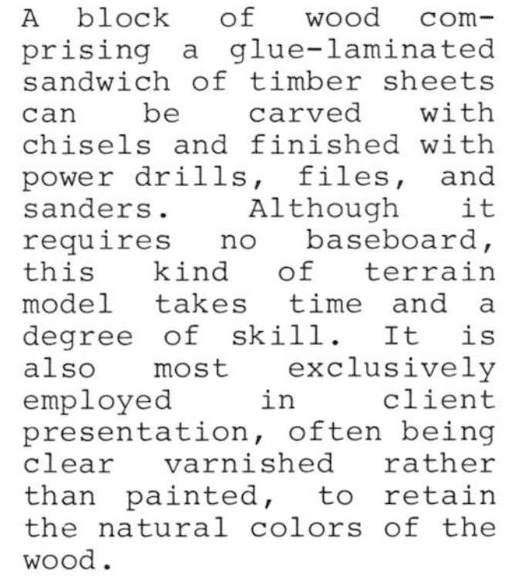

4

Methods of Rendering the Terrain

1

Both the technique of application and the choice of coloring medium in modelmaking depend upon the model type and its surface material. For example, if a model presents a single building, a common approach is to leave the terrain in rugged textural contrast to the precision of the simulated built form. This contrast is especially attractive in models using chipboard, plastic foam, and corrugated cardboard for the terrain. It is heightened by adding dark pigments to the setting, against the light color of the proposed building. The converse of this approach is to paint or spray the terrain in white or a light color. This is common on town-planning scale models where richer colors may be confined to the proposed building as a means of creating a center of attention.

Paints should be carefully blended to a consistency that will not distort the surface of the terrain. Paints that are excessively thinned in water or white spirit can warp paper, cardboard, and wood surfaces.

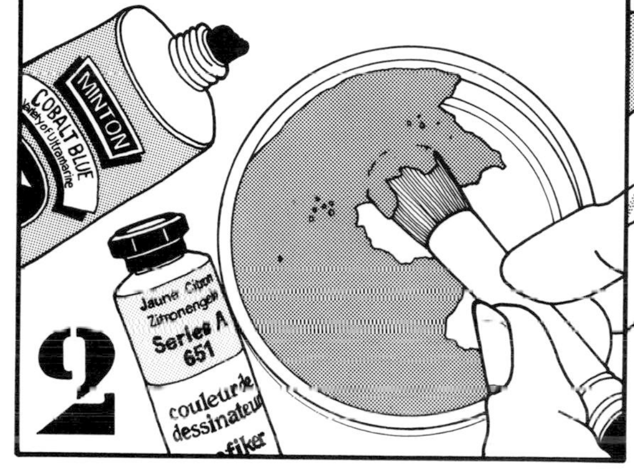

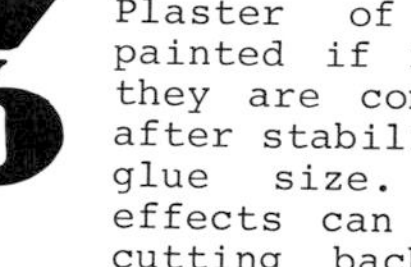

Plaster of Paris models can be painted if required, but only when they are completely dry and often after stabilizing their surfaces with glue size. Interesting landscape effects can be achieved by lightly cutting back through the painted surface using sandpaper.

4

As a general rule, water-based pigments only should be applied on foam plastics and Styrofoam. Other paints, as with adhesives, should be tested before application, as many attack and dissolve synthetic materials. Also, when acrylic sheet is used in modelmaking, this requires a spray-applied nitrogen resin priming lacquer.

5

If required, contours in paper and wood can be softened with a thin coat of emulsion paint or smoothed down with plaster of Paris or a water-based crack filler medium.

The terrain can then be painted or laminated with a variety of glue-applied finishes, including felts, sandpapers, earth, and so on.

6

Water- and oil-based color-spray treatments of terrain can be fast and effective, the atomizing of the pigment bringing the possibility of scaled surface effects.

A rapid method of rendering the terrain is to spray single- or multi-color "washes" through cutout paper masks to establish road systems, areas of landscape, and the like.

How to Mix Plaster of Paris

1

Pour two or three pints of water into a plastic bowl and add plaster of Paris (or dental plaster) by dropping handfuls into the water.

N.B.: The dropping action reduces the incidence of bubbles in the subsequent mix.

2

When the plaster of Paris forms a peak just above the waterline, the contents of the bowl are ready for mixing.

3

To mix, plunge a hand to the lower part of the bowl and agitate rapidly with the flat of the hand. When thoroughly mixed, check its consistency by inspecting the mixture on the back of the hand. If it appears to be an opaque cream, it is ready for use.

4

Hardening time is rapid and generates heat, so it must be applied quickly. For this reason several mixes may be required, applied as built-up layers.

N.B.: The setting time can be slowed down by the addition of a small quantity of size (or beer) to the water prior to mixing.

PLASTER OF PARIS

SIZE

5

Methods of application range from pouring and troweling to hand flicking. The latter should be used when producing large-scale terrain as it reduces the incidence of bubbles in the mix.

6

When required, areas of plaster of Paris can be reinforced, by dipping strips of sacking into the mix and laminating onto the model during the buildup process.

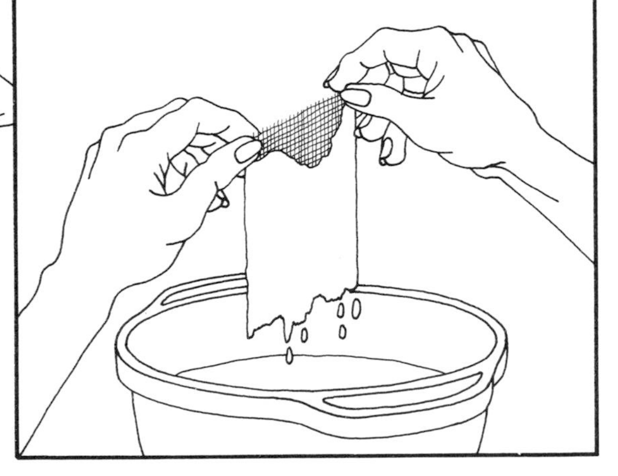

How to Make Papier-mâché

1

Tear a large quantity of newspaper into small pieces, the smaller the better. Place in a bucket, cover with water, and leave to soak.

When thoroughly soaked, work to a pulp by squeezing with the hands. During this process, pour away surplus water at regular intervals and replenish with clean water. When pulped, drain off surplus water.

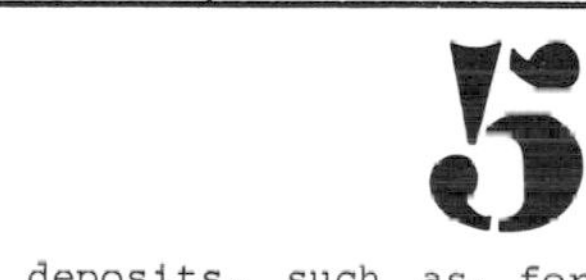

2

Next, add a water-based colorant such as white powder color or emulsion paint, together with a wallpaper paste mixed with water according to the manufacturer's instructions. When this mixture is readily hand molded, it is ready for use.

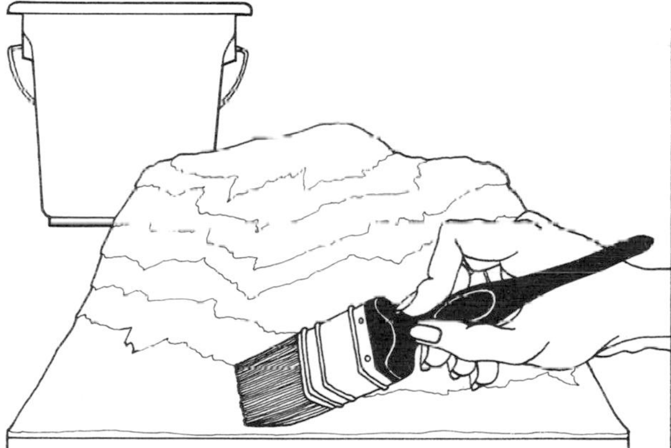

Papier-mâché is applied in built-up layers. These can each be smoothed by brushing over with wallpaper paste or hot glue size. This application is important over extra-thin layers and especially around the edges of a modeled terrain so that it adheres to the baseboard.

4

5

Avoid extra-thick deposits, such as, for instance, those exceeding 4" (100mm) thickness, as these rapidly exhaust the supply of papier-mâché and generate moisture that can warp baseboards.

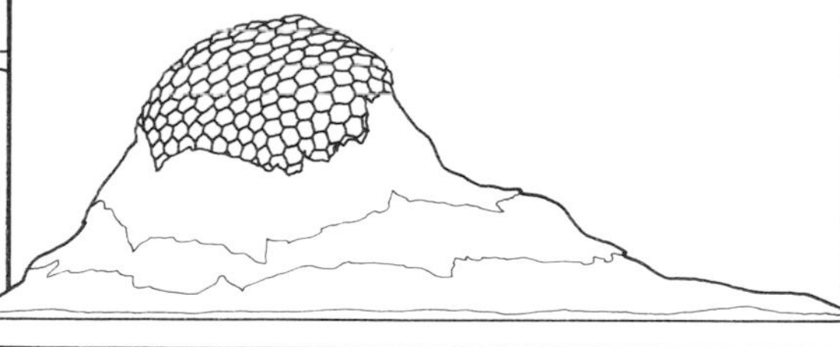

Mounds can easily be supported by armatures of wire mesh, crumpled paper, or cardboard (see page 106).

6

When the modeling process is finished, papier-mâché should be left to dry thoroughly and naturally. When dry, any shrinkage cracks can be stopped with a filler, excavations can be made, the surface can be sanded, and, if so required, brush painted or color sprayed.

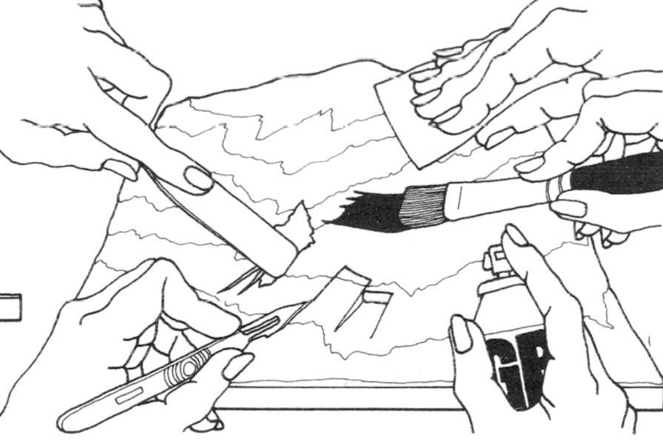

Basic Paper Folding and Assembly Technique

Paper is the quickest and most economical means from which to produce an endless variety of sharp-edged architectural forms, particularly in the construction of small- to medium-scale block models of all types.

Quick-setting, all-purpose, and colorless model-making adhesives aid rapid assembly, adjoining edges being hand held while the glue sets.

N.B.: Adhesives are discussed fully on page 97 in Manual 1.

2 When crisp, clean joints are required, it is wise to include gluing tabs during the precutting stage. These may be placed on any one side of two edges to be joined.

Prior to gluing, all corners and fold lines along the glue tabs should be scored along their outside edges with the point of a compass or knife against a straightedge. They should then be prefolded and creased with a finger so that the intended form is readily assumed during assembly.

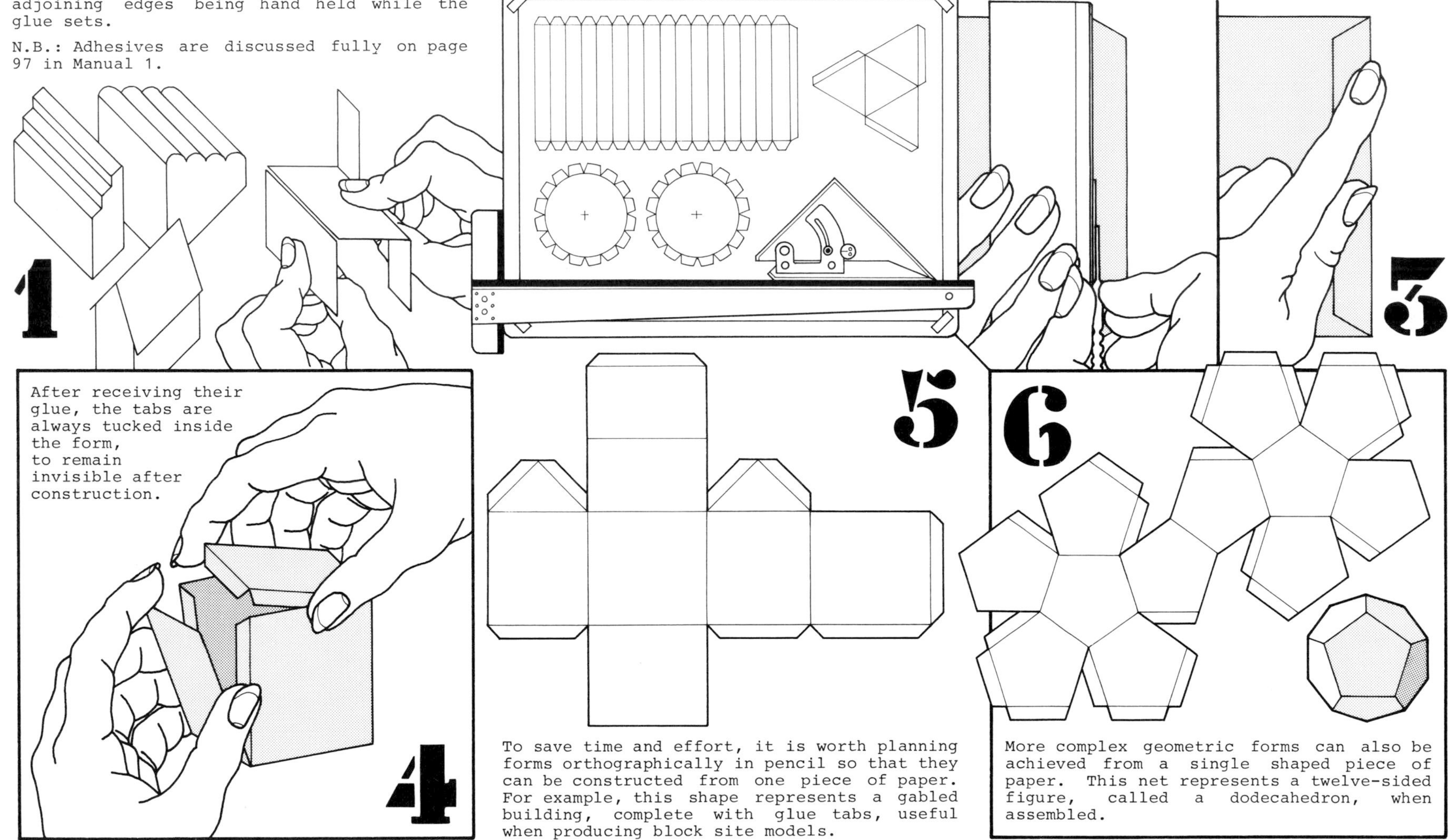

After receiving their glue, the tabs are always tucked inside the form, to remain invisible after construction.

To save time and effort, it is worth planning forms orthographically in pencil so that they can be constructed from one piece of paper. For example, this shape represents a gabled building, complete with glue tabs, useful when producing block site models.

More complex geometric forms can also be achieved from a single shaped piece of paper. This net represents a twelve-sided figure, called a dodecahedron, when assembled.

Buildings in Card and Balsa Wood

1 When building hollow-construction models of designs, first transfer their dimensions from the orthographic drawings to the required scale onto the card or balsa-wood sheet.

N.B.: Card-mounted diazo-printed orthographics can be cut out and assembled into fast, effective models.

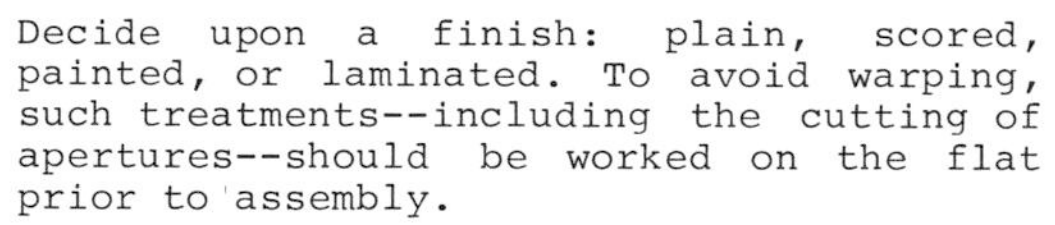
Decide upon a finish: plain, scored, painted, or laminated. To avoid warping, such treatments--including the cutting of apertures--should be worked on the flat prior to assembly.

Cut out the basic components of the model, always cutting with the surface finish uppermost. Corners can be butt jointed or mitered for a more professional finish.

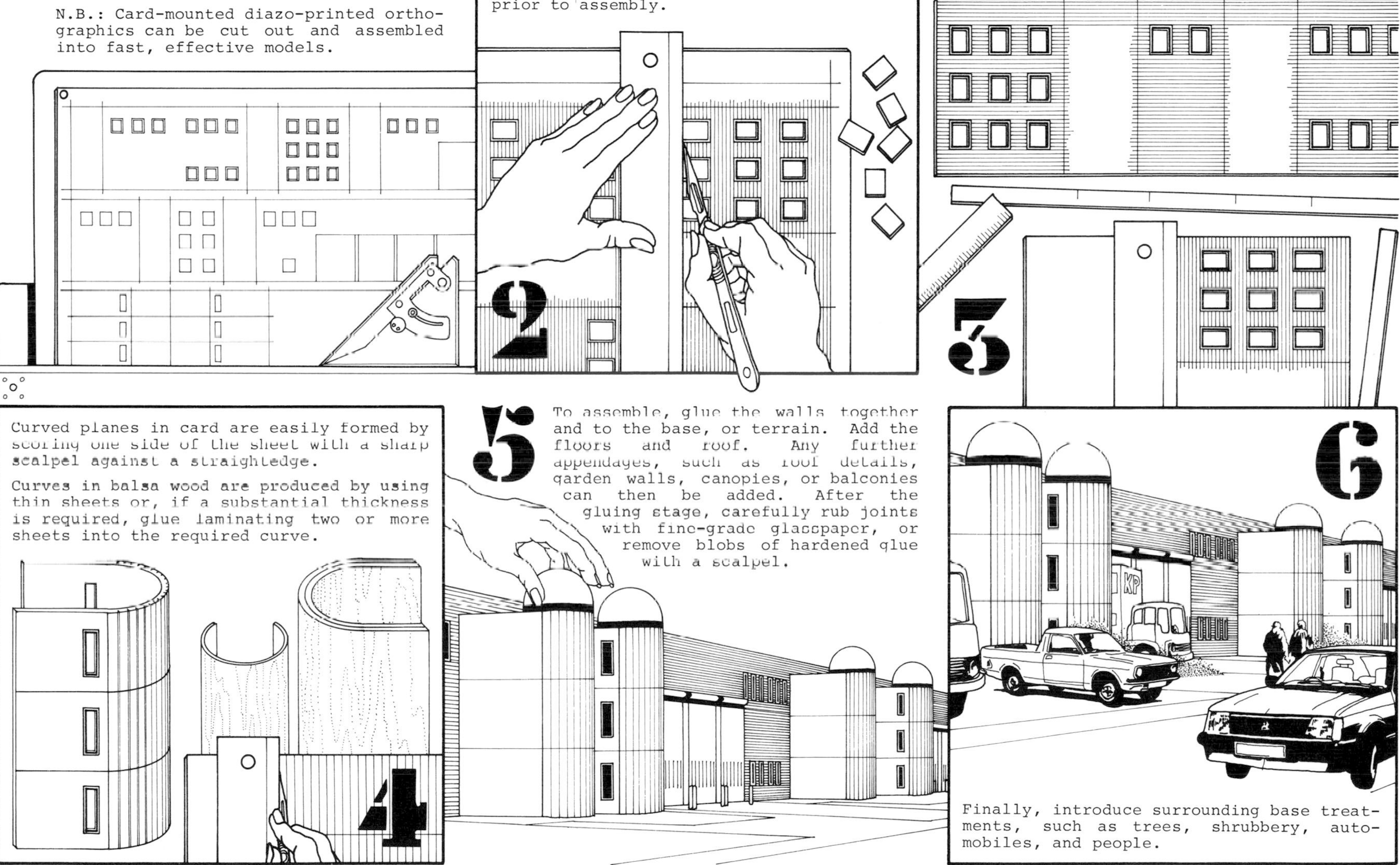

Curved planes in card are easily formed by scoring one side of the sheet with a sharp scalpel against a straightedge.

Curves in balsa wood are produced by using thin sheets or, if a substantial thickness is required, glue laminating two or more sheets into the required curve.

To assemble, glue the walls together and to the base, or terrain. Add the floors and roof. Any further appendages, such as roof details, garden walls, canopies, or balconies can then be added. After the gluing stage, carefully rub joints with fine-grade glasspaper, or remove blobs of hardened glue with a scalpel.

Finally, introduce surrounding base treatments, such as trees, shrubbery, automobiles, and people.

How to Model Trees and Shrubbery

1

Natural sources for tree and shrubbery simulation include twigs, heather, dried herbs (such as yarrow), fir cones, pine needles, lichen, sponge, loofah, woodshavings, etc. These can be used as found, or brush painted, color sprayed, or colored by soaking in pigment.

2

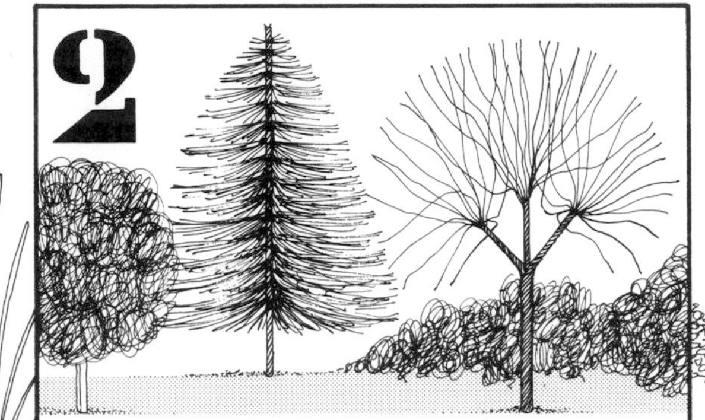

Natural-looking trees from man-made materials include coiled electric copper wire, strands being splayed to simulate branches above a coiled section that represents the trunk; steel wool glued directly to the terrain or supported on trunks made up of matches, pushpins, or toothpicks, and bottle brushes, pruned to simulate cypresses.

3

Deciduous trees can be heavily endowed with foliage by dipping armatures of splayed wire or yarrow into glue before a second dipping in a dried mixture of sawdust and color.

N.B.: The sawdust-color mix is also good for application to the terrain as "instant grass" or groundcover.

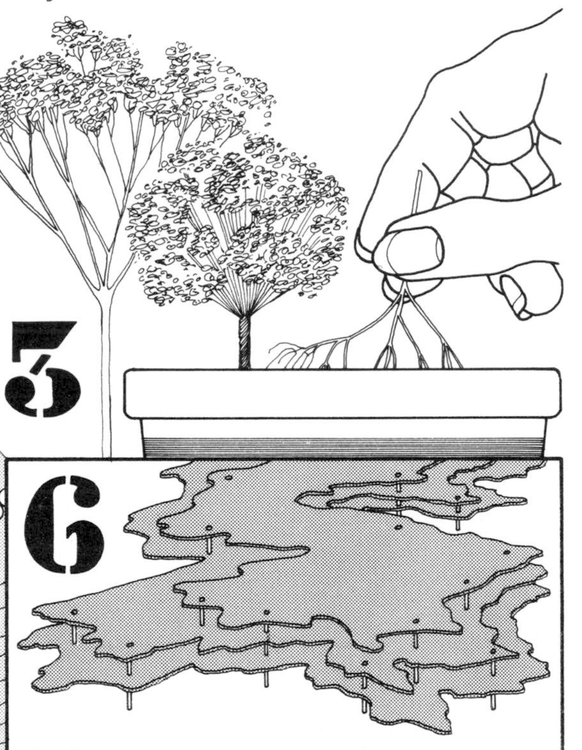

4

A common method of simulating foliage is to use plastic or sponge rubber, sculptured or chopped into shape with a scalpel or scissors. If required, it can then be color sprayed before being applied directly, supported on "trunks," or cut into strips to represent hedgerows.

N.B.: Well manicured hedgerows can be cut out from a soft, green composition block used as a support for flower arranging, and purchased in florists.

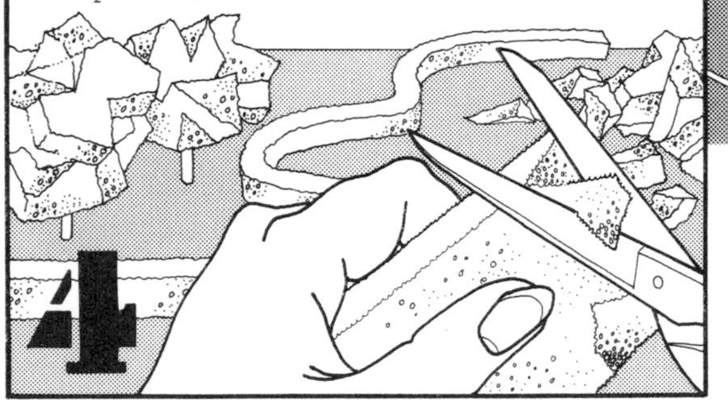

5

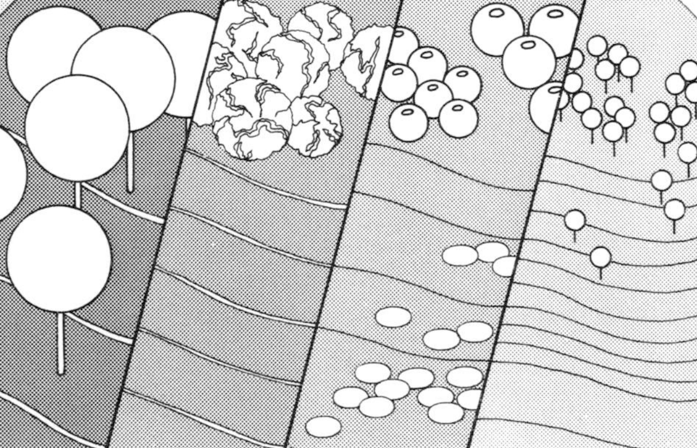

Depending on their scale, schematic models attract trees from unusual sources. These include table tennis balls (effective on all-white models), various sizes of wooden and plastic beads, dried peas, pushpins, map tacks, and crushed balls of thin paper in colors sympathetic to the model.

6

A fast method of producing schematic forests and clumps of trees is to cut a series of three or four horizontal sections in cardboard that describe their collective mass. These are then supported by toothpicks or matches that pierce their layers.

N.B.: Trees tend to be highly vulnerable on finished models. Therefore, whenever possible, it is wise to drill holes into the terrain to receive their glued supports.

How to Simulate People, Automobiles, and Water

1 The inclusion of simulated human figures in models is a means of identifying size and scale during both the design and communication stages.

At larger scales, figures can be bent from wire, carved from balsa wood, sculptured in clay, or cut from card and bent into action positions. For models to be photographed, card-mounted photographs of people cut from magazines, or cut-out silhouettes in black or gray card of people traced from magazines, are extremely convincing.

As scale descends, however, so the level of abstraction is increased, and diminutive figures are symbolized rather than replicated.

N.B.: Crowds of people in models should be grouped to suggest movement, or clustered as features of interest. Individual figures are better located near such other scale-giving elements as windows, columns, or railings.

2 In order to avoid the distraction that a faithful toy-shop replica can bring to models, automobiles are best modeled in abstract form. They can be carved from Styrofoam or balsa wood and colored to suit the overall quality of the model.

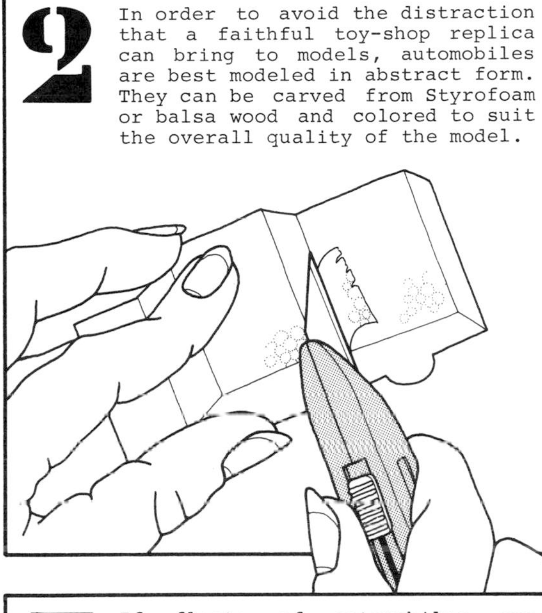

3 If fleets of automobiles are required, these can be cast in plaster of Paris, liquid plastic, or candle wax from a rubber mold. The mold is made by brushing layers of latex medium over the wooden master. When set, the mold is flexible and easily peeled away.

N.B.: Rubber molds can, if necessary, be turned inside out between casts and cleaned with soap and water.

4 Areas of water are best represented by sheer, translucent, transparent, or reflective materials, or by a pigment or colored surface that exploits some contrast within the finish of the model.

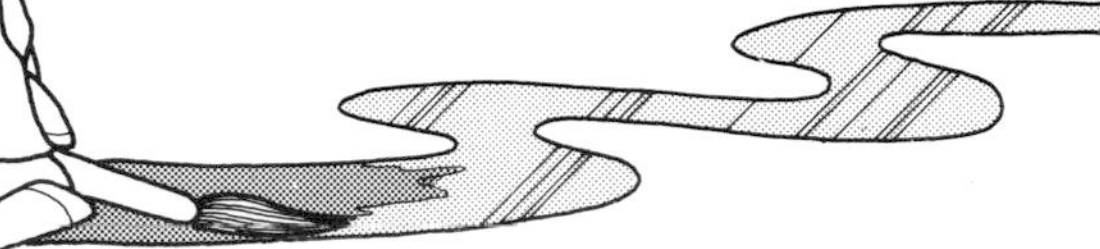

Paint under clear varnish or a satin or glossy finish paper or card.

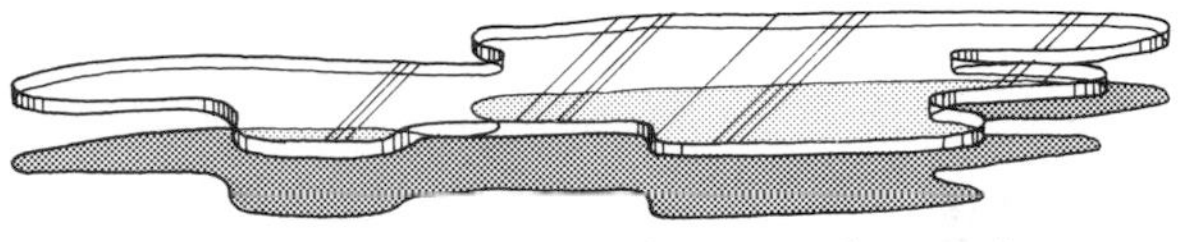

Blue Plexiglas. Self-adhesive dry color, or paper, sandwiched under clear acrylic, acetate film, or glass.

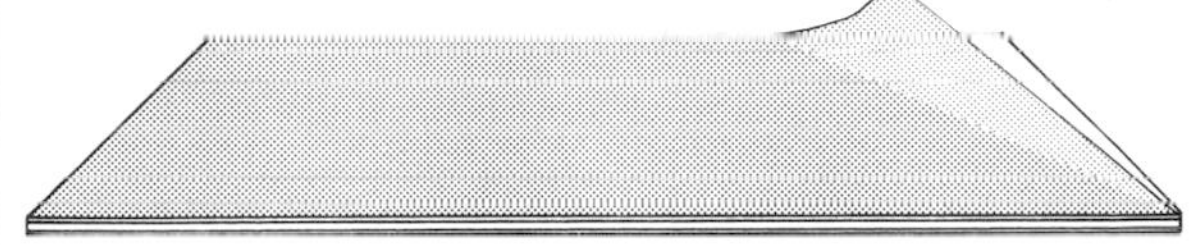

Transparent blue self-adhesive color sheet laminated on white card.

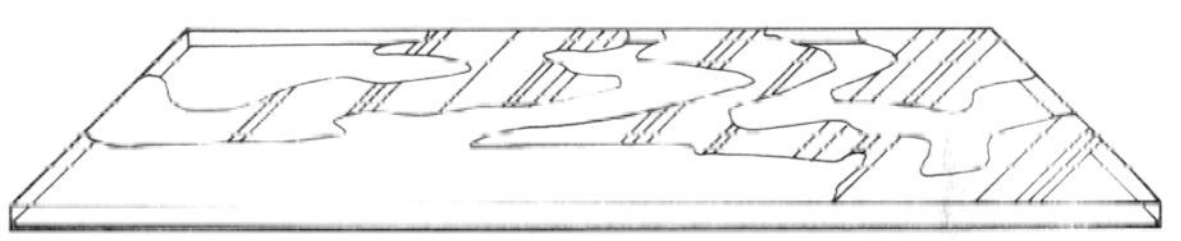

Mirror glass, aluminum paper, or foil applied to exploit a "rippled" finish.

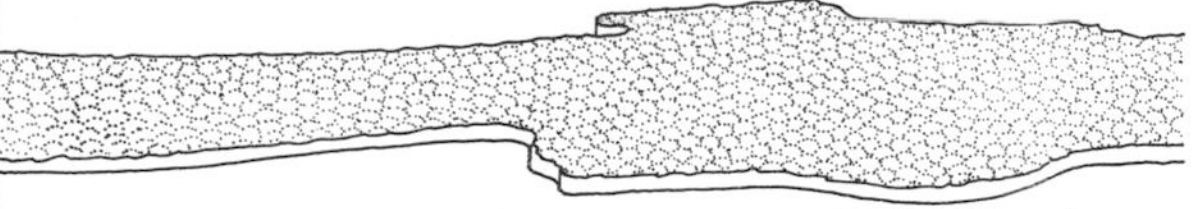

Ripple-texture glass, or cathedral glass, appears remarkably effective when used in large areas on town planning or site models at larger scales.

N.B.: Areas of water in models can usually be secured on the baseboard by its surrounding terrain. Otherwise, pins, adhesives, or clips can be concealed at points where land overlaps water.

Tips when Elaborating Models: Visual Cues

Architectural scale involves the relationship of people to the size of their environment. The cues that enable viewers to sense this relationship are thus important elements to include in the production of models.

1

In the vertical plane there are numerous devices that can be introduced into models that establish the important sense of scale, as they are roughly based on the dimensions of anthropometry. These include repetitive elements, such as mortar joints in building materials, door sizes, balustrades, patterns of fenestration, mullions, columns, and so on.

Colors also carry with them a strong sense of scale. For example, a brilliant green or a bright red, used respectively to represent grass or brick, would appear too intense.

One should apply a built-in sense of scale when mixing colors for models. These should generally be desaturated, that is, grayed down to adjust to the model's scale. First test samples on a scrap of paper before introducing them into the model.

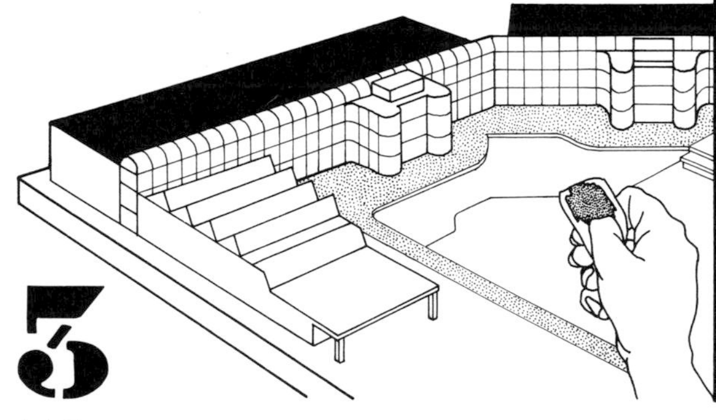

3

4

The appearance of brushmarks on painted small-scale models can also destroy their sense of scale. For this reason, extra-special care should be taken when color rendering to ensure that the grain of application is compatible with the grain of scale.

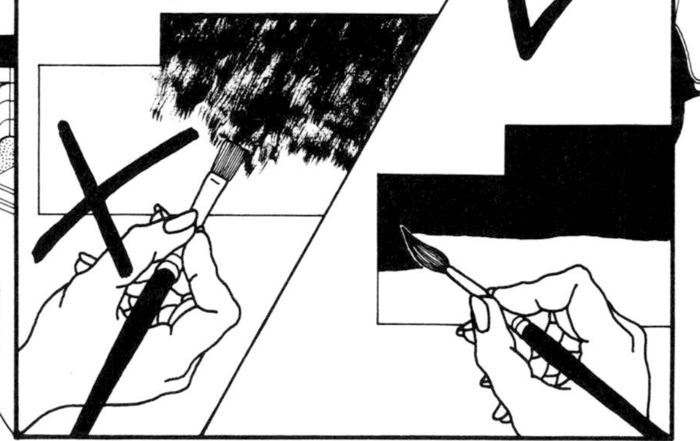

2

Familiar objects and floorscape treatments on the horizontal plane also serve as strong scale indicators, as they tend to fall into categories of size. These include sidewalks, curbs, paving patterns, steps, parking slots, signs, automobiles, trees, and, of course, human figures.

N.B.: Although many custom model-making papers are available for simulating surfaces, such as brickwork and grass, they should be used with caution, because their printed regularity can overwork the sense of scale and lead to a "toy" effect.

Tips when Elaborating Models: Visual Cues

5

Any study of the built environment provides many hints to aid in the model-making process that aims to simulate its appearance. For example, when structuring value in models, roofs generally appear darker than walls--even when they are inclined to the light--because of shadows cast from overlapping elements. Glazing and foliage also tend to "read" as dark elements and should be organized as such in naturalistic models.

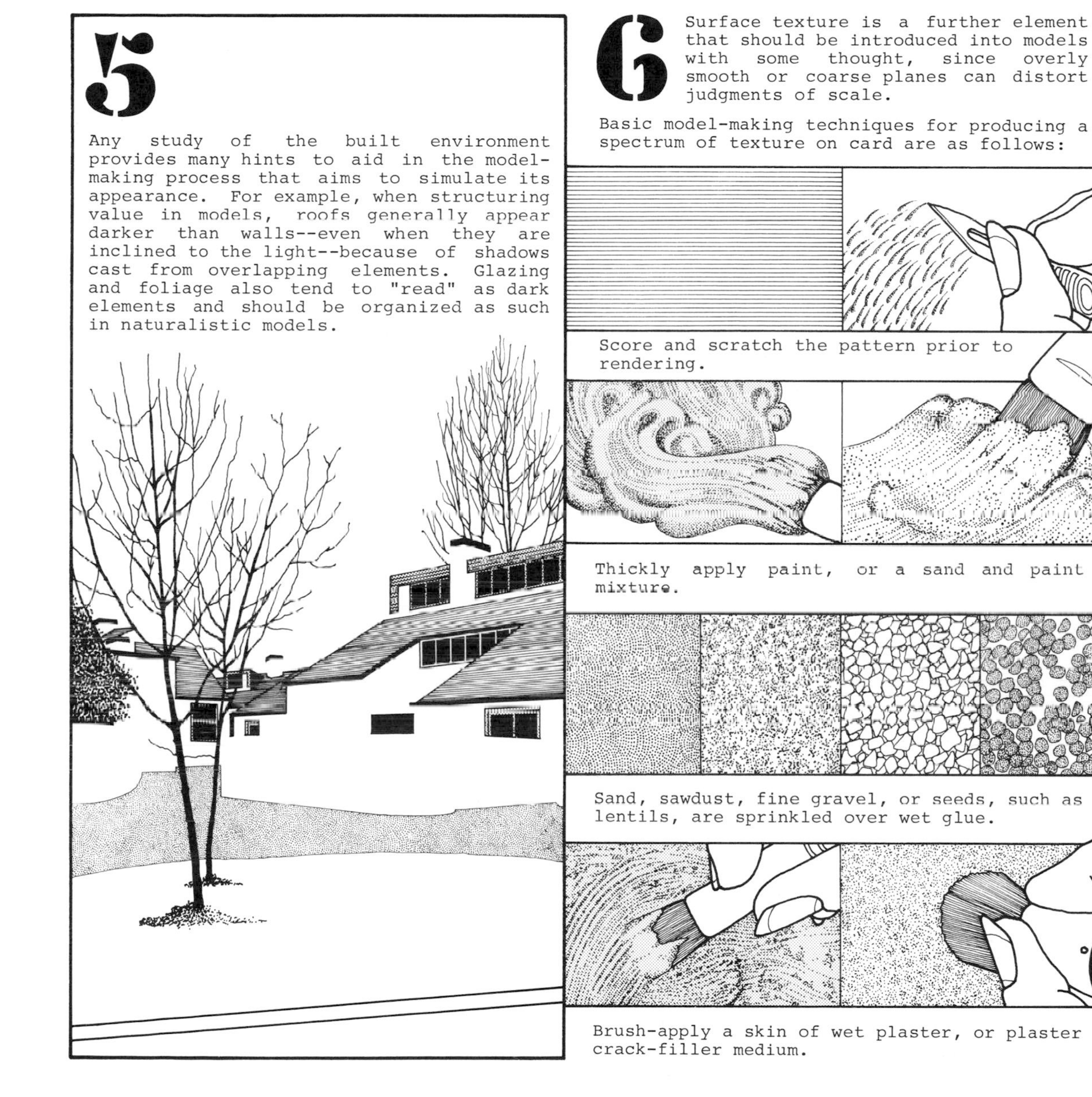

6

Surface texture is a further element that should be introduced into models with some thought, since overly smooth or coarse planes can distort judgments of scale.

Basic model-making techniques for producing a spectrum of texture on card are as follows:

Score and scratch the pattern prior to rendering.

Thickly apply paint, or a sand and paint mixture.

Sand, sawdust, fine gravel, or seeds, such as lentils, are sprinkled over wet glue.

Brush-apply a skin of wet plaster, or plaster crack-filler medium.

7

Textures can also be achieved by laminating various grades of abrasive papers, such as glasspaper and carborundum paper, onto card. These are produced in a useful range of earth colors, grays and black being ideal for roofs and roads, ochres and yellows for gravel paths and concrete. Abrasive papers can also be color sprayed to simulate grass, earth, stucco, stone, etc., in model surfaces at a larger scale.

Tips when Making Site Models Using Photographs

1

Urban design and site models built using scaled photographic prints of buildings taken in the actual site and then laminated onto the surface of recipient cardboard forms provide highly realistic settings in which to design or present modeled architectural proposals. They also provide rich mini-environments against which building proposals can be photographed, and highly convincing "sets" for video films that simulate movement toward and around a designed form.

N.B.: The following hints are based on the use of a shift lens that corrects the perspective distortion experienced with normal lenses (see page 120).

2

A good strategy is to methodically preplan the sequence of shooting the area that is to be represented in the model on a site plan. This is used during site photography, each shot being marked off on the plan to ensure a comprehensive survey.

N.B.: To avoid confusion during processing, always shoot façades in one continuous direction.

3

One method of achieving consistency of scale is to shoot each building from a set distance from its façade. The appropriate model scale is later fixed on the enlarger easel, using the first negative. This setting is then maintained for the enlargement of all subsequent negatives.

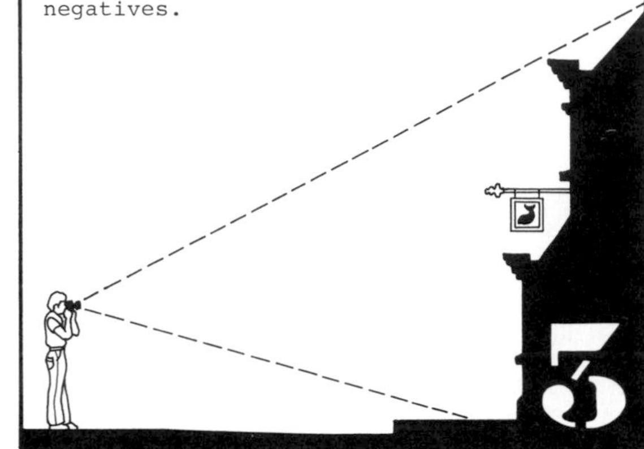

4

As the maintenance of a fixed distance from façades is not always possible, another method is recommended.

This involves the use of a roll of white and a roll of black adhesive tape. Prior to shooting, a small strip is attached to each façade (black tape for light-colored buildings, white for dark) 6 feet (1800 mm) above sidewalk level.

5

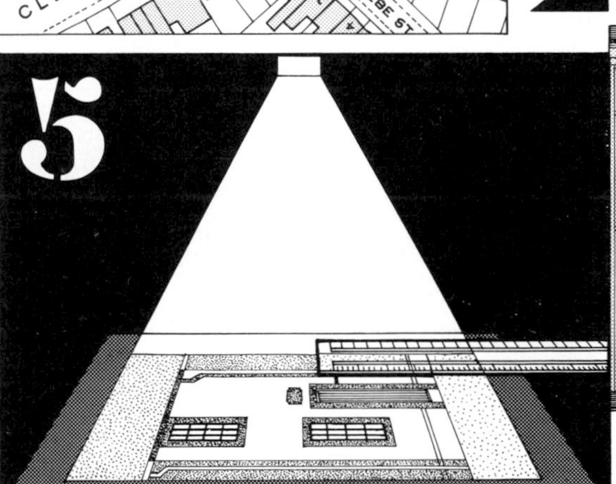

Later, in the darkroom, these strips will be detectable on the negative image on the enlarger easel and, with the aid of a scale rule, will determine and maintain an accurate scale reading throughout processing.

6

A timesaving and economical way around the problem of shooting long, repetitive façades, or those blocked by parked autos and the like, is to shoot one representative section. This can then be printed to scale as right way around and mirrored images to the number required in order to achieve a composite montage of the complete elevation.

Tips when Making Site Models Using Photographs

7

When shooting excessively long façades as a series of photographs, again, shoot continuously. Also, as the slightest camera movement can cause distortion of the image and, as edge distortion can occur in perfect shooting conditions, always allow a generous margin around the façade being shot. A good tip is to work within the middle third of the viewfinder image, the margin of overlap being later trimmed away from the print.

N.B.: In order to achieve a consistent value level in panoramic and other composite prints, always develop such related shots together.

Trees, hedges, streetlights, and the like, which may occur directly in front of façades, need not pose a problem for models built from photographs, especially if their modeled replicas are to be introduced. The effect in the finished model is simply that of a three-dimensional form with its photographed version occurring on the façade immediately behind it.

8

9

To summarize, a good sequence is to first establish the scale of the model and, working from an accurate site plan, photograph the site area before building the model. At the model-making stage, two kinds of construction can be considered. The first is a simple "stage set" model that describes only the planes of street façades.

10

The other is a complete block model that represents all the planes of the buildings.

N.B.: The smallest advisable scale for models using photographs is 1/16" = 1' 0" (1 : 200).

11

As both methods will involve corner joints, a good assembly method of achieving precision joints is to first trim the cardboard into the required components. Then laminate the prints so that, at each corner, one print is flush with the card and the other overlaps by the thickness of the card. When glue assembled, the result is a crisp corner.

12

In order to bring something of the reality of the façades to the floorscape, make sure that mortar joints in sidewalks and road markings are introduced into the model. The latter can be inserted by applying cut strips of appropriately colored masking tape.

N.B.: Roof planes in complete block models are best represented in dark or black card.

Models Using Photographs: The Shift Lens

1

When photographing tall building façades, vertical lines are kept parallel by holding the film plane of the camera parallel to the building. However, with a normal lens the resultant photograph tends to include too much foreground while excluding the top of the building. If the camera is tilted skyward to include the whole façade, vertical lines begin to converge, making the building appear to distort and lean backward.

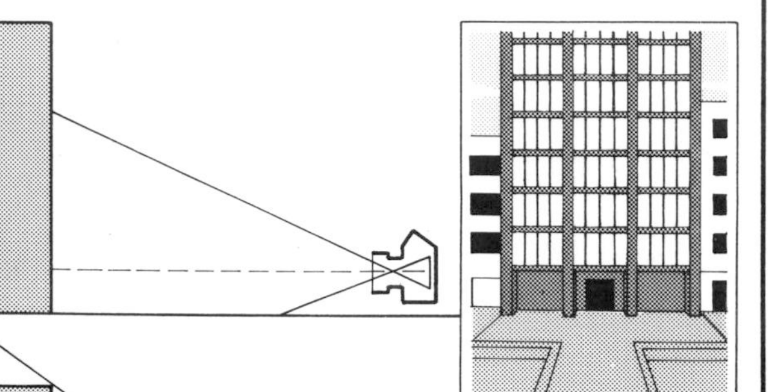

2

While a certain degree of distortion can be corrected during processing by tilting the enlarger easel or by later cropping of unwanted foreground portions, these methods are tedious and are simply overcome by the use of a perspective control, or shift lens

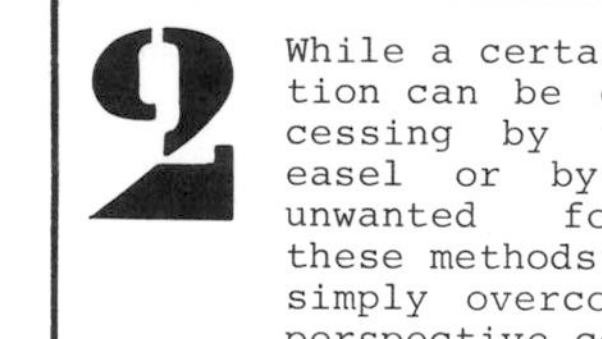

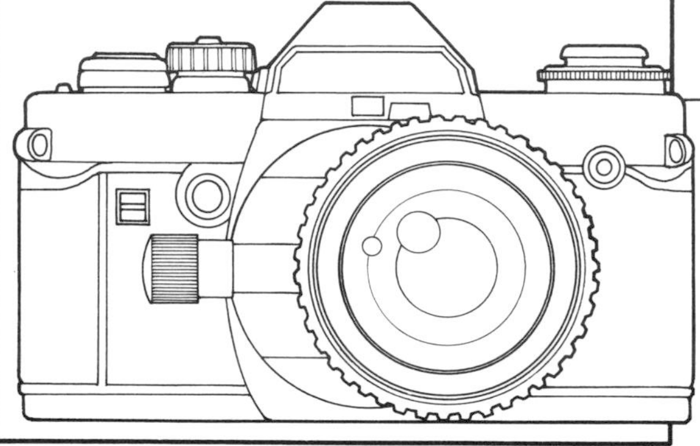

3

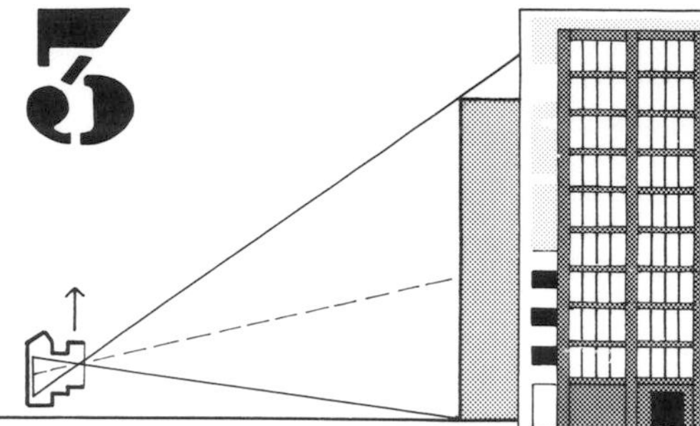

The shift capability of perspective control lenses allows most converging or diverging parallels to be corrected. This is done simply by keeping the camera's film plane parallel to the building and, when the verticals are established, shifting the lens upward until the entire building fits within the picture frame.

N.B.: Some shift lenses incorporate a focusing screen etched with a grid to help align linear objects.

4

Although the shift lens is obviously useful for shooting tall buildings in corrected perspective, it also has horizontal capabilities. For example, when you are required to shoot a building, but cannot place yourself directly in front of it, or when buildings are fronted with unwanted objects such as a tree or a traffic light, get as close to the center as you can and turn the lens so that it shifts toward the center of the building. Next, turn the control knob until it appears in the viewfinder as if you were actually standing directly in front.

5

When joining two pictures together--as will often be called for when shooting photographs for models--any directional shift in the camera's shooting angle will produce unjoinable center lines that cause the two parts of the panorama to mismatch.

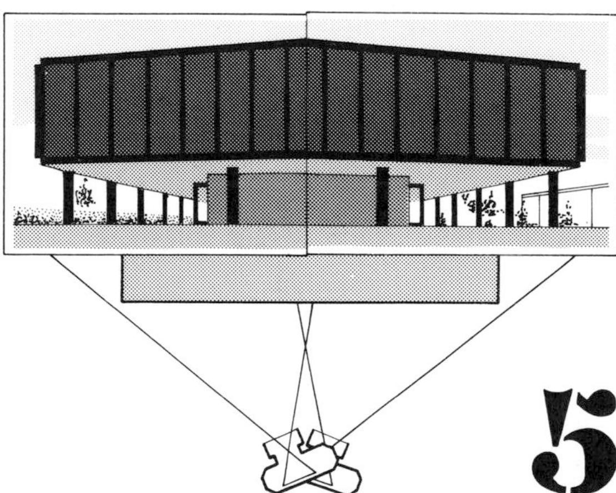

6

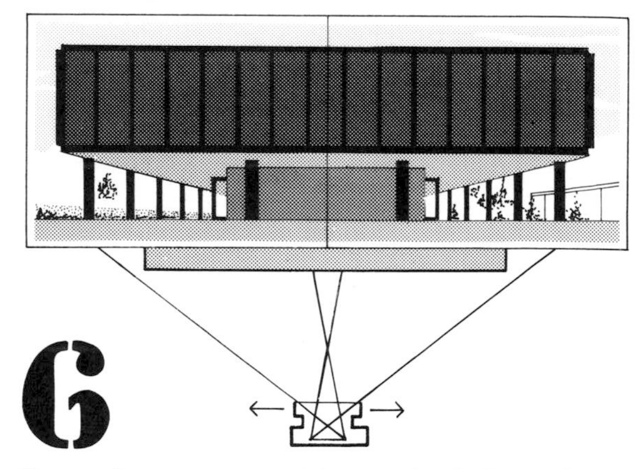

When using a perspective control lens, however, the camera is kept stationary while the lens is shifted in each direction between shots to provide perfectly corrected and matching images.

N.B.: Such shots involve overlap. Therefore, the prints require trimming to achieve the composite picture.

A Textured Finish for Site Models

1 The antithesis of a realistic site model constructed from laminated photographs of existing buildings is a block model built quickly from found scraps of paper and board.

2 However, a good method of unifying its appearance is to introduce an all-over textured finish. To do this, first mix some sand with an emulsion paint of the color of your choice.

3 Once thoroughly mixed, this is then brush applied over the whole surface of the model.

N.B.: Make the mix as dry as possible. One application is enough to coat the surface.

4 This is a site model of Gloucester Green, Oxford, including a mixed development building proposal treated with the sand-paint texture by Ben Enwonwu and Richard Hutchings, two fifth-year architecture students at Oxford Polytechnic. In order to produce a passive coloration an off-white emulsion paint was used.

Modelmaking Kits

1 Proprietary modelmaking kits are on the market that can provide a ready-made source of components for rapid assembly of town planning scale models and those that plan the layouts of offices and industrial plants.

One such kit, developed from a children's plug-in construction toy, comprises a range of basic shaped plastic components from which models of any scale can be built.

2 The basic interchangeable components can be assembled to symbolize furniture, equipment, people, and so on, at larger scales such as 1 : 50. The kit also comprises standard window, door, wall, and beam units.

3 Models are built up by plugging the components onto a pegged baseboard of gray plastic, the area of the model being extended by the addition of further baseboards.

N.B.: Transparent acrylic sheets are also supplied for transferring of information by tracing from drawings into the model.

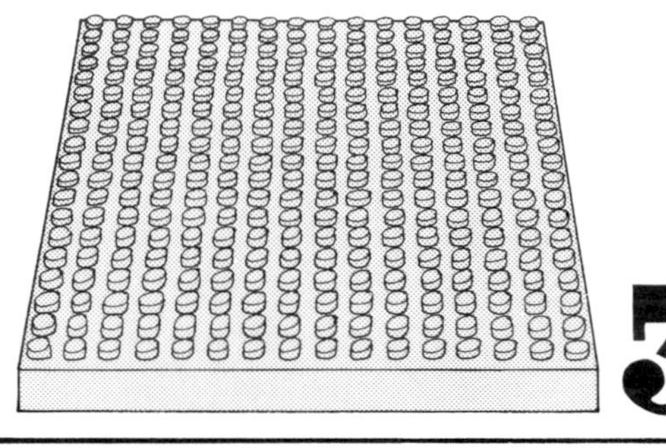

4 The components are self-colored so that, when grouped, they play a role in communicating color-coded information, such as service and circulation routes, departmental zones, phases of long-term development, and so on.

The baseboard can also be colored by overlaying custom plastic perforated sheets that are trimmed to size to plan parking lots, roads, footpaths, etc.

5 Together with do-it-yourself models, these ready-made modeling systems force the designer to confront spatial and structural problems that may remain unnoticed in drawings. By presenting an uncompromising toylike quality, their central function is limited to that of space planning. They also invite the important participation in decision making of interested parties outside the design team.

Photographing Models Against the Site

Designers who photograph their models in the open air in natural light help make things easier for themselves, as models look best in front of natural sky and clouds. Professor A. C. Hardy at the University of Newcastle devised an ingenious method of producing a composite photograph of a model against its intended setting as part of his research into the effects of the color of farm buildings on the countryside.

1

So that the model can be elevated to the eye level of a camera set on a tripod, the baseboard is supported on a thin wire bracket attached to a wooden pole. The pole is then tacked or wired to a stake driven into the ground, which allows vertical on-site adjustment.

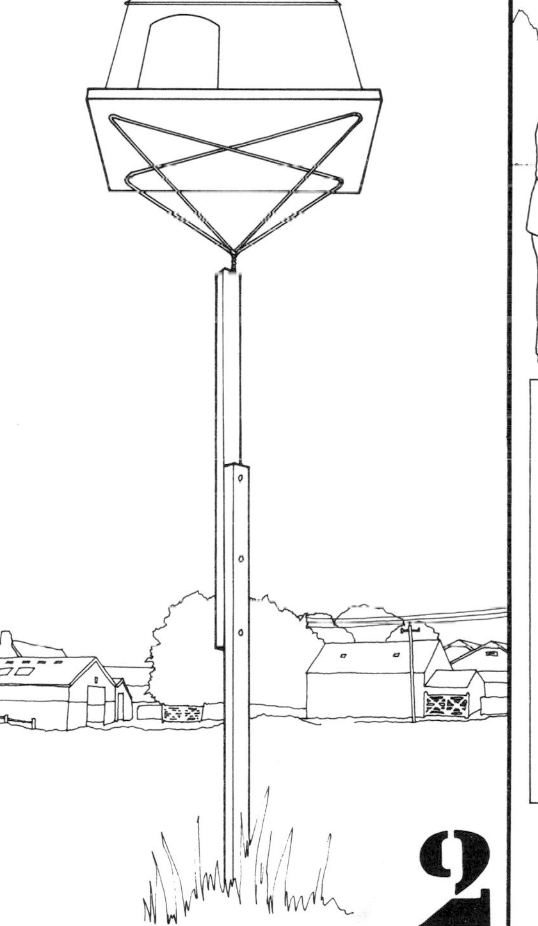

2

3

The model is then fixed in its elevated position at some distance from the site itself. The exact distance is determined carefully by the point at which its relative size and scale appears compatible with that of its background setting through the camera viewfinder.

In order to compensate for the focal discrepancy between the foreground model and the background view, a macro lens is used to photograph the composite image.

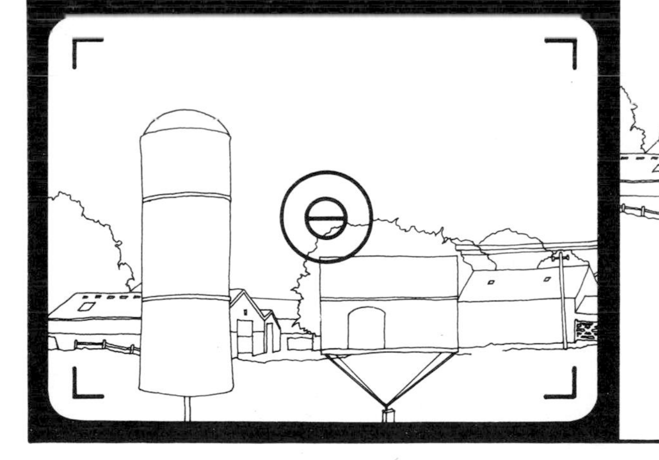

Although this method is governed by the type of model used and, indeed, by the long-view visual accessibility of the site, it can provide stunning and convincing photographs.

Modelscope Photography: Composite Image Technique

Achieving a photograph of an architectural model that will give an impression of how the project will ultimately appear to an observer standing in its precincts can present a number of problems. For example, the camera is usually too bulky to be accommodated within the confines of the model, its focal axis being difficult to position at the scale of simulated eye level. Also, normal lenses do not provide an adequate depth of field in such short-focus situations.

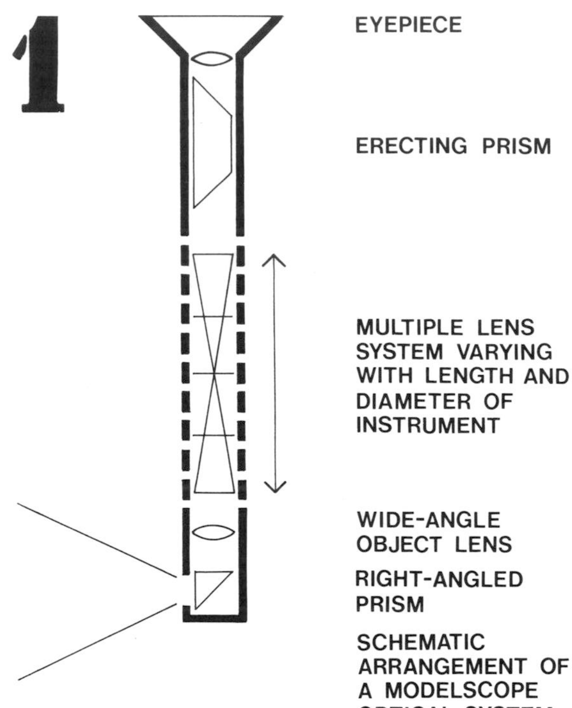

These problems can be overcome by attaching a modelscope, or periscope, to the camera. However, modelscope photography can also introduce its own difficulties. These include poor light transmission combined with picture distortion and a falling-off in sharpness toward the edge of the circular pictures. The following tips and techniques therefore are offered as a means of achieving the best possible results.

2

When purchasing a modelscope instrument for photographic use, evidence of its capabilities should always be obtained in the form of test pictures. Here is such a test picture. The numbers represent distances in centimeters from the viewing point.

4

It is important to note that the angle of field, that is, the amount of information viewed, covered in a picture is determined by the optical chatacteristics of the modelscope. Alternative lenses in the camera will only spread or diminish the size of image produced on the film. Although modelscope manufacturers often provide long-focus (telephoto) lenses to achieve greater coverage of the normal picture format, this is achieved only at the expense of image brightness, necessitating excessively long exposure times. Modelscopes, however, provide a very simple means of achieving wide-angle coverage. Richard Abbott has devised an interesting technique comprising three or more overlapping exposures that can be trimmed into a rectangular image. Although originally designed for his Execuscope instrument, the technique is applicable to other types of modelscope.

Modelscopes coupled directly to the front of a camera will normally project onto the film a circular image with a diameter equal to one-fifth the focal length of the camera lens. Thus, a 35mm camera with a standard 50mm lens will give a .4 inch (10mm) diameter image. If fine-grain photographic materials are used, the basic image will enlarge comfortably by ten diameters.

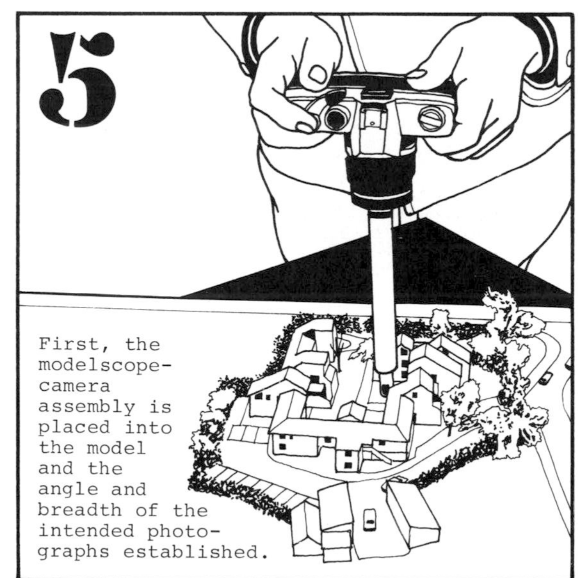

First, the modelscope-camera assembly is placed into the model and the angle and breadth of the intended photographs established.

Modelscope Photography: Composite Image Technique

A cardboard platform graduated in 25-degree intervals is next placed into the model at the point at which the modelscope head will be positioned for shooting. Each gradation in turn will control the precise angle of each shot in the sequence.

N.B.: The gradations are positioned behind the modelscope head, each angle being aligned to a mark on the back of the modelscope head before each shot.

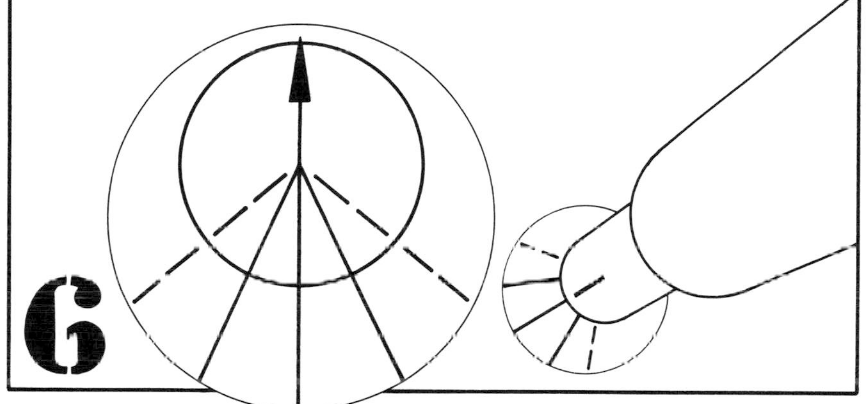

6

Now, take the first shot and rotate the modelscope 25 degrees after each exposure. If the camera is tripod mounted, movement between exposures can usually be made by rotating the modelscope in its mounting, without disturbing the camera. If the camera is hand held, a check on the vertical alignment of the modelscope before each shot can be made with a small spirit level placed on the back of the camera.

7

8

Provided that lighting levels, camera settings, and film processing remain constant, little difficulty will be experienced in matching the three individual prints that, allowing for overlap, will provide a field of view of approximately 90 degrees. The three exposures also provide a "wide screen" impression of a far greater reality than do single exposures and can give a composite processed picture of up to 12.6 x 9.8 inches (320 x 250mm).

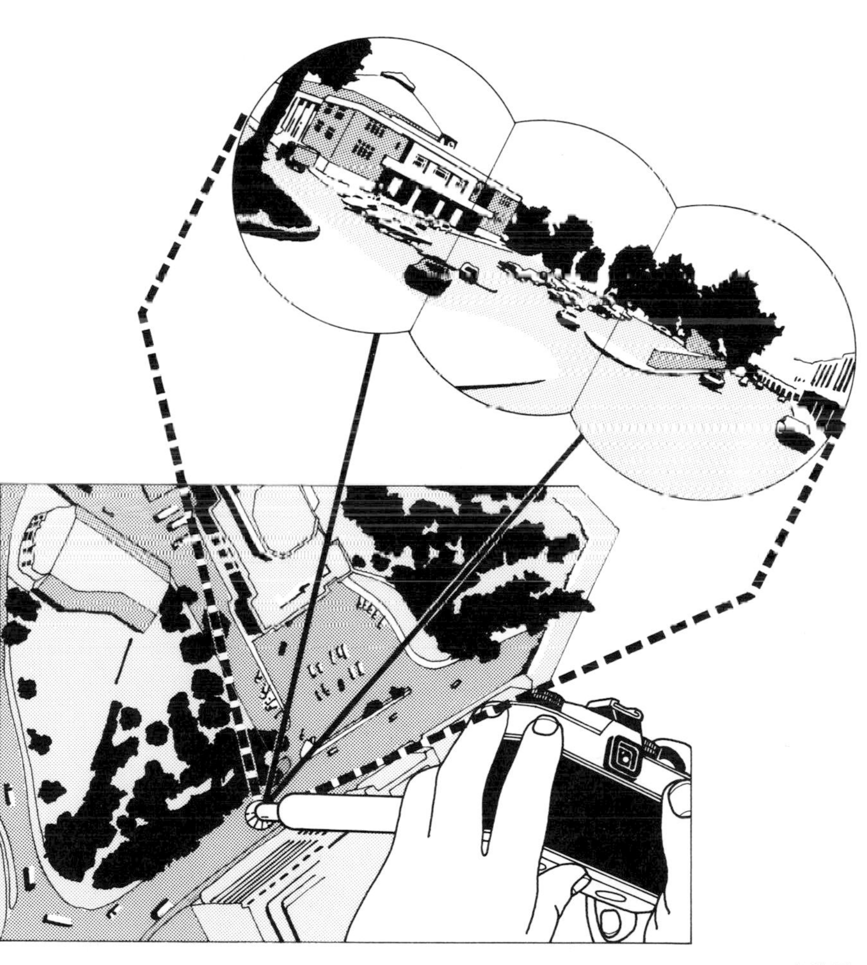

Modelscopes: Photographic Lighting Techniques

1

Photographing models outdoors, either with or without a modelscope attachment, offers two advantages. First, it may be possible to provide an appropriate natural background of foliage and sky, if not that of the actual intended site. Second, natural lighting will be uniform and possibly bright enough to permit a short exposure.

Strong direct sunlight causes heavy shadows and should either be avoided or counterbalanced by a reflector, photofloods, or flash. A do-it-yourself reflector can be simply a sheet of white card positioned to reflect sunlight into shadowed areas on the model.

2

3

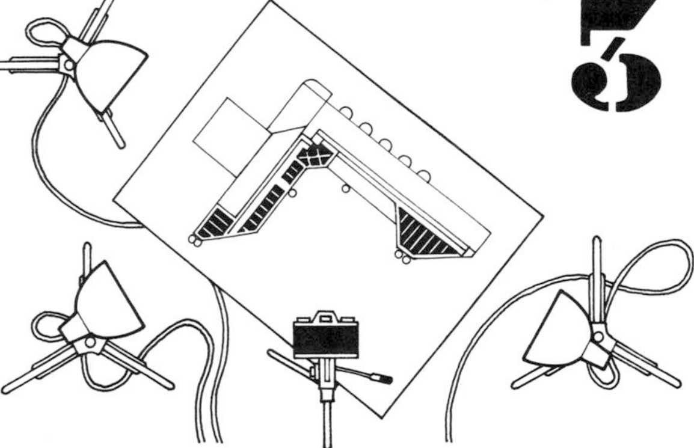

Photofloods provide indoor illumination. However, illumination diminishes in proportion to the square of the distance from the light source, so the position and number of lamps is critical. For example, lamps positioned alongside the viewing point will often require supplementary lamps forward of this position in order to fill in near areas of the model.

A total of three or four lamps may be needed, the whole area to be photographed being checked section by section with a light meter to ensure uniformity. As they are highly combustible, it is vital that spirit- or petroleum-based glues not be used in models subjected to the heat of lamps.

N.B.: Indoor modelscope photographs can also be taken by firing a flash upward into a white reflector umbrella (or a white ceiling) held directly over the model. A synchronized slave flash fired from one side will give a shadow effect. This is quite a convenient way of lighting modelscope shots but, unlike photoflood or natural lighting, the effect is not seen until the prints are developed.

4

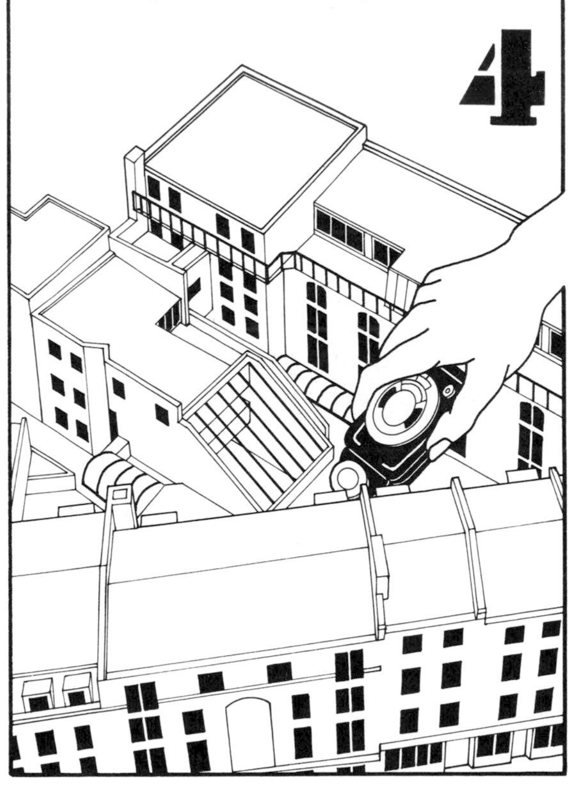

Modelscopes: Composite Triple Exposures

1

These are two examples of the triple-exposure technique described on pages 124-25. The photographs, taken by Richard Abbott of a 1 : 100 scale model (top) and a 1 : 200 scale model (bottom), illustrate how his technique can be used on both large and small projects.

N.B.: When required, a fourth exposure can be made to extend the width of the field of view.

Both of these triple-exposure photographs were achieved in the following manner. The three circular modelscope pictures were first trimmed carefully to achieve an accurate joining into a composite panorama, then dry-mounted in position on a paper backing sheet. The skyline profile was then trace-cut with a sharp knife to remove the background behind the model.

Next, a photographic print of a natural backdrop, that is, of a sky or treescape (taken especially for this role), was inserted above and behind the skyline. The montage of four prints was finally cropped into the rectangular format, in readiness for reshooting as a second-stage print.

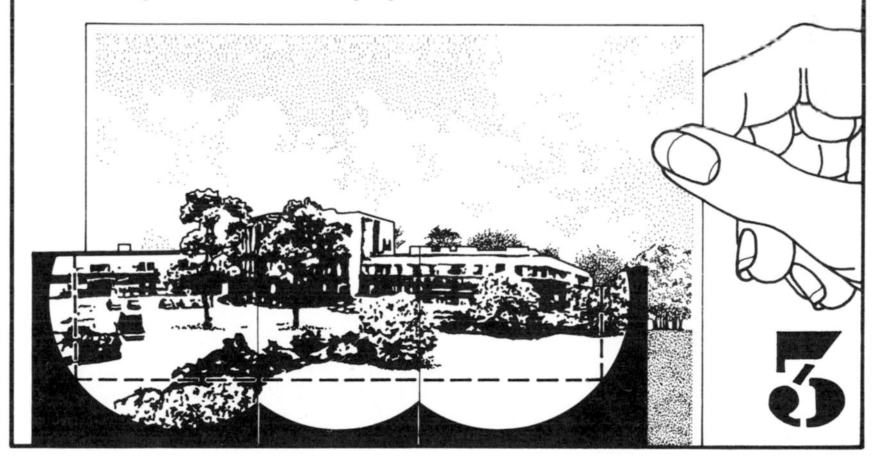

INDEX